Materials Forming, Machining and Tribology

Series editor

J. Paulo Davim, Aveiro, Portugal

More information about this series at http://www.springer.com/series/11181

Yuriy Batygin · Marina Barbashova
Oleh Sabokar

Electromagnetic Metal Forming for Advanced Processing Technologies

Springer

Yuriy Batygin
Kharkiv National Automobile
and Highway University
Kharkiv
Ukraine

Oleh Sabokar
Kharkiv National Automobile
and Highway University
Kharkiv
Ukraine

Marina Barbashova
Kharkiv National Automobile
and Highway University
Kharkiv
Ukraine

ISSN 2195-0911 ISSN 2195-092X (electronic)
Materials Forming, Machining and Tribology
ISBN 978-3-030-09024-1 ISBN 978-3-319-74570-1 (eBook)
https://doi.org/10.1007/978-3-319-74570-1

Printed on acid-free paper

This Springer imprint is published by the registered company Springer International Publishing AG part of Springer Nature
The registered company address is: Gewerbestrasse 11, 6330 Cham, Switzerland

Contents

Chapter 1
Introduction

The practical use of the energy of pulsed electromagnetic fields opens exclusive perspectives for the creation of advanced technologies for the processing of materials of any physical nature. In special terminology, these technologies are termed "magnetic pulsed metal working out" (MPMW [Russian references]) or "Electromagnetic Metal Forming" (EMF [Western references]). These technologies have doubtless advantages, namely, ecological purity, low energy consumption, economic expenditure of the material resources, and high speed of the manufacture process. In addition, it is necessary to mention a very important fact connected with the appearance of new metal alloys possessing unique properties, which become apparent only under pulsed action, for example, the hyper-plasticity that occurs when the relative deformations are under pulsed loading may reach approximately 200%.

As was pointed out already, electromagnetic metal forming involves an impulse or high-speed forming technology that uses pulsed electromagnetic fields to apply forces to tubular or sheet-metal work-pieces. A recent detailed review by Psyk et al. (2011) provided a historical perspective of the development of EMF processes and highlighted the state of the art on modeling, coil design, sheet-metal formation, tube formation, crimping, welding, cutting, calibration of parts, and hybrid processes involving EMF. It should be noted that in a lot of unique literature sources, for example, Belyy et al. (1977), high specific conductivity of both the metal of the work-piece and the metal of inductor is the major requirement for the effective realization of EMF processes, which can be considered "traditional" at present.

Despite the "traditionality" of magnetic pulsed processing of well-conducting metals, many applications exist for creating new advanced technologies with the use of powerful electromagnetic field energy. For example, the production operation of angle formation in sheet metals may lie in the foundation of such a technology.

As follows from phenomenological considerations, the mentioned technological operation becomes possible when the pressure forces are concentrated in the centre of the angle formed. In any case, that is why one of the most important questions in

Y. Batygin et al., *Electromagnetic Metal Forming for Advanced Processing Technologies*, Materials Forming, Machining and Tribology,
https://doi.org/10.1007/978-3-319-74570-1_1

designing tools for magnetic pulsed methods is the one about the aim-directed force action on the billet area, which must be deformed in accordance with the production order.

Here some new suggestions for the construction of the needed tool are necessary. They must agree with the analysis of electrical dynamic processes in the system "inductor–billet" with subsequent recommendations for their practical fulfilling and installation.

In the present work, several constructions of an inductor system are offered on the basis of considering the use of excited electromagnetic fields. Their space–time distribution allows passing to the real construction based on satisfying the requirements of strictly directed force acting on the centre of the angle being formed.

Concretely, the works described here were initiated by the problem of forming sharp angles in stamping the back panel of an automobile that is being issued by the Ford Motor Company and shown on Fig. 1.1.

The initial creation of the required configuration of the panel was produced by known electrohydraulic methods. However, creating the angle with the clearness level being demanded according to the production task turned out to be impossible in practice. In this regard, the application of magnetic pulsed technology in a regime involving the hyper-plasticity of metal treatment was developed.

Constructively, the suggested and investigated inductor systems are represented by combinations of the simplest single-turn solenoids of rectangular shape. Physically the given combinations first provides a rather high level of electro-dynamic interaction of the excited magnetic field and the current being induced in the billet metal.

The above-mentioned application of EMF is based on *repelling* by Lorentz forces, which are excited between the coil-tool and the conductive blank. In addition, to the repelling processes of the billet, another configuration of EMF processes is possible when the latter is *attracted* to the coil.

Interest in the magnetic-pulse attraction of thin-walled conductors has a long-standing history. Some different suggestions allowing artificial transforming of the natural magnetic pressure in the process of magnetic attraction appeared during the first tests of some field technologies. The practical side of the marked interest

Fig. 1.1 The sheet billet of an experimental sample of the back panel of a car body

was initiated by the growing demand on production operations by the effective restoration of airplane and vehicle bodies with dents. The process must be realized from the body external side. This must be flattened without disassembling and without damaging the existing protective cover. EMF technologies precisely satisfy to all listed requirements.

The next objective of this work is representing some theoretical and experimental results for a new manufacturing process of sheet-metal attraction through the use of recently developed inductor systems. This process is one of the few suggestions regarding the generation of attracting forces with much higher capabilities than vacuum formation.

The first application of the attraction process is expected to take place in automotive repair shops where an attracting force may enable the correction of panel damage from the exterior side.

A final remark: The material of this book is being provided without concrete references of the original sources because the authors consider such a method of exposition more comfortable for reading. The full bibliography can be found in the final pages of the book.

References

Psyk, V., Risch, D., Kinsey, B. L., Tekkayaa, A. E., & Kleiner, M. (2011). Electromagnetic forming—A review. *Journal of Materials Processing Technology, 211*, 787–829 (Elsevier).

Belyy, I. V., Fertik, S. M.,& Khimenko, L. T. (1977). Spravochnik Po Magnitnoimpul'Snoy Obrabotke Metallov [Electromagnetic Metal Forming Handbook]. English translation by Altynova, M.M., online available at http://www.mse.eng.ohiostate.edu/_Daehn/metalforminghb/index.html (shown on 04.11.2010).

Chapter 2
Magnetic Pulsed Pressure for Forming Inner Angles in Sheet Metals

The distribution of internal angular surfaces is a competitive technology for processing metals by pressure and can be used as an element of the technological cycle for the production of various elements of metal structures. This paragraph presents the choice of optimal configurations of the executive tool-inductor by direct solution of the problem of calculating the induced currents and the arising mechanical forces. The problem is based on the Maxwell equations of integration for geometric and electrophysical parameters, which have been given according to the standard using classical mathematical apparatus. The optimal arrangement, which is mathematically obtained for the inductor current circuits, leads to the maximum concentration of forces generated at the center of the angle and causes harmful side pressure on the metal of the sides of the structure.

Realizing some technological operations in the flat stamping of sheet metals requires concentrating forces of pressure for the "filling in of angles" according to designed shape of a ready article. The term "filling in of angles" is a special term derived from the forging and stamping production. In practice, it means creating strict-enough angles with the rounding-off level being designed by the production-operation conditions.

As has been shown in scientific works, the action of a high-speed pulse leads to appearance of a new behavioral property of the processed metal. This new property, as described by G. S. Dahen, Ph.D., is termed "hyper-plasticity." Its practical usage allows the formation of with the help of the known mechanical methods without destroying the metal.

Magnetic pulsed metalworking in practice realizes the power force–acting regime in the time interval with the duration of a microsecond. The main particularity of such processing methods is the indissoluble connection of the inductor-tool and the work-piece that is the subject of deformation. If, during the mechanical stamping, its efficiency is not connected directly with the inner processes in the processed metal, the action of the magnetic pulse force is physically possible only when an electromagnetic connection exists between the inductor-tool coil current and the current that is induced in the work-piece. Their interaction leads

Y. Batygin et al., *Electromagnetic Metal Forming for Advanced Processing Technologies*, Materials Forming, Machining and Tribology,
https://doi.org/10.1007/978-3-319-74570-1_2

to the appearance of high-power ponder-motor forces, the value of which is proportional to produce these current amplitudes.

2.1 Problem Formulation Under Idealization of Inductor Current Wires by Thin Wires

One of the main problems in designing tools for magnetic-pulse methods is the aid-directed force acting on the work-piece area, which must be worked out in accordance with the production order. As was mentioned previously, analysis of the electrical magnetic processes in the system termed "inductor–work-piece," with the following recommendations for their constructive production and inter-arrangement, is necessary.

The aim of the present consideration is calculation of the fields' space–time distribution and the electrical dynamic forces being excited in the inductor system where the described forces are acting on the inner cavity of the flat angle in the crook of the metal sheet with quite high specific electrical conductivity.

It should be noted that an identical problem was described in the science monograph of G. A. Shneerson, Sc.D.

The inductor is long single-turn solenoid of rectangular shape. One of its sides faces the angle "being filled in" as shown in Fig. 2.1, which shows a system cross-section. In addition, the calculation model is adopted.

For solving the formulated problem, some following statements must be accepted.

- The angle in the crook area of the flat metal sheet is close to approximately $\frac{\pi}{2}$.
- The Decatur's rectangular coordinate system is adopted.

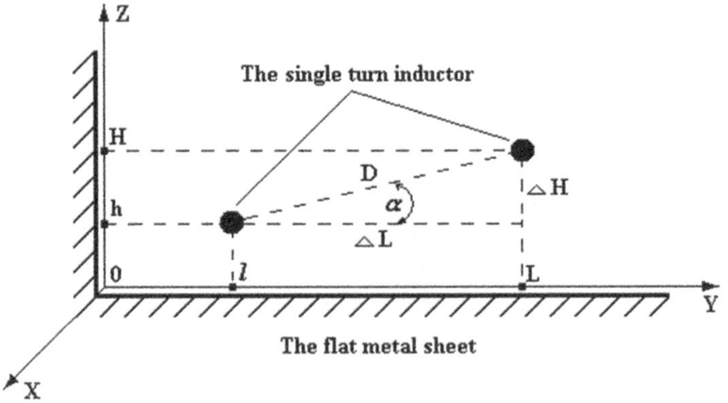

Fig. 2.1 Cross-section of a schematic model of the inductor system with a single-turn rectangular solenoid in an inner zone of the angle of the bent metal sheet

- The single-turn solenoid of the rectangular shape is made from a thin conductor and it is long in direction of OX-axis so that $\frac{\partial}{\partial x} = 0$.
- The specific electrical conductivity of the processed metal is quite high; thus, the calculation may be conducted with help of the known condition of the skin-effect.
- Through the current wires of the turn, which are parallel to the OX-axis, the same current is flowing $I_x(t) = I\ (t)$ (t is the time in). Its direction in the current wire with coordinates (l, h) coincides with the positive direction of the OX-axis, but it is reversed in the current wire with coordinates (L, H) it is reverse.

The adopted statements allow to suggest that in the system being considered the following electrical magnetic field intensity vector components will be excited as follows: $E_x(t,y,z) \neq 0,\ H_y(t,y,z) \neq 0,\ H_z(t,y,z) \neq 0$.

As it follows from statement about skin-effect on the metal sheet surfaces, the electrical field intensity tangent component will be equal to zero, that is, $E_x(t,y,z=0) = E_x(t,y=0,z) = 0$.

Solving the formulated problem will be conducted identically as has been done in the known publications in which theoretical investigations of inductor systems with thin-walled conductors were represented.

2.2 Integrating Maxwell Equations and Calculating Relationships

The Maxwell equations for the electrical magnetic field intensity of non-zero components, Laplace-transformed with the zero initial conditions in the space between the inter-orthogonal planes of the metal sheet ($y \geq 0,\ z \geq 0$), are as follows:

$$\begin{cases} \dfrac{\partial H_z(p,y,z)}{\partial y} - \dfrac{\partial H_y(p,y,z)}{\partial z} = j_x(p,y,z), \\ \dfrac{\partial E_x(p,y,z)}{\partial z} = -p\mu_0 H_y(p,y,z), \\ \dfrac{\partial E_x(p,y,z)}{\partial y} = p\mu_0 H_z(p,y,z), \end{cases} \qquad (2.1, 2.2, 2.3)$$

where p is the integral Laplace-transforming parameter;

$j_x(p,y,z)$—is the current density in the current wire;
$j_x(p,y,z) = I(p) \cdot [\delta(z-h) \cdot \delta(y-\ell) - \delta(z-H) \cdot \delta(y-L)], I(p) = L\{I(t)\}$,
$\delta(z), \delta(z)$—are the Dirac's delta functions; and
$E_x(p,y,z) = L\{E_x(t,y,z)\}, H_{y,z}(p,y,z) = L\{H_{y,z}(t,y,z)\}$.

From the differential system (2.1) through (2.3), the equation for $E_x(p, y, z)$ may be this:

$$\frac{\partial^2 E_x(p,y,z)}{\partial y^2} + \frac{\partial^2 E_x(p,y,z)}{\partial z^2} = p\mu_0 j_x(p,y,z). \tag{2.4}$$

For solving Eq. (2.4), the Fourier sinus transformation may be used. A choice of the given mathematical approach is caused by the boundary condition.

Thus, we have

$$E_x(p,y,z) = \int_0^{\infty} E_x(t,\lambda,z) \cdot \sin(\lambda y) \cdot d\lambda, \tag{2.5}$$

$$j_x(p,y,z) = \int_0^{\infty} j_x(t,\lambda,z) \cdot \sin(\lambda y) \cdot d\lambda, \tag{2.6}$$

where

$$j_x(p,\lambda,z) = \frac{2}{\pi}\int_0^{\infty} j_x(t,y,z)\sin(\lambda y)dy = I(p)\cdot(f_1(\lambda)\cdot\delta(z-h) - f_2(\lambda)\cdot\delta(z-H),$$

$$f_1(\lambda) = \frac{2}{\pi}\cdot\sin(\lambda\ell),\ f_2(\lambda) = \frac{2}{\pi}\cdot\sin(\lambda L).$$

Taking into account formulas (2.5) and (2.6), Eq. (2.4) may be transformed into the following:

$$\frac{\partial^2 E_x(p,\lambda,z)}{\partial z^2} - \lambda^2 E_x(p,\lambda,z) = K_1(p,\lambda)\cdot\delta(z-h) - K_2(p,\lambda)\cdot\delta(z-H), \tag{2.7}$$

where $K_1(p,\lambda) = \mu_0 p \cdot I(p) \cdot f_1(\lambda)$, $K_2(p,\lambda) = \mu_0 p \cdot I(p) \cdot f_2(\lambda)$.

Equation (2.7) is the usual linear differential equation of the second order with the constant coefficients with the right part, which is represented by the delta-functions linear combination.

The superposition principle may be applied.

At the beginning, the solution of the equation, the right part of which contains the algebraic summand—$F_1(p,\lambda,z) = K_1(p,\lambda)\cdot\delta(z-h)$, should be found. It will be the function—$E_x^{(1)}(p,\lambda,z)$. Then the second, which is the function—$E_x^{(2)}(p,\lambda,z)$, may be found for the right part $F_2(p,\lambda,z) = -K_2(p,\lambda)\cdot\delta(z-H)$. The sum of the obtained results $E_x(p,\lambda,z) = E_x^{(1)}(p,\lambda,z) + E_x^{(2)}(p,\lambda,z)$ will give the equation solution for the electrical field intensity satisfying the boundary conditions in the skin-effect regime.

The general integral of Eq. (2.7) with the right part $F_1(p,\lambda,z)$, which was found by the Lagrange-method, appears as follows:

$$E_x^{(1)}(p,\lambda,z) = C_1(p,\lambda)\cdot e^{\lambda z} + C_2(p,\lambda)\cdot e^{-\lambda z} + \frac{K_1(p,\lambda)}{\lambda}\cdot \eta(z-h)\cdot \mathrm{sh}(\lambda(z-h)), \tag{2.8}$$

where $C_{1,2}(p,\lambda)$—are the constants of integration, $\eta\ (z)$—is the Heaviside step function.

The constants of integration may be defined from the boundary conditions for the intensity on the horizontal plane, $E_x^{(1)}(p,\lambda,z=0)=0$ as well as the limitation conditions in the infinity, $E_x^{(1)}(p,\lambda,z\to 0)=0$.

After necessary mathematical transforming and according substituting, we may obtain the following:

$$E_x^{(1)}(p,\lambda,z) = 2\frac{\mu_0 pI(p)}{\pi\lambda}\cdot \sin(\lambda\ell)\cdot\left[\eta(z-h)\cdot \mathrm{sh}(\lambda(z-h)) - e^{-\lambda h}\mathrm{sh}(\lambda z)\right]. \tag{2.9}$$

The second solution of differential equation (2.7) is defined identically:

$$E_x^{(2)}(p,\lambda,z) = -\ 2\frac{\mu_0 pI(p)}{\pi\lambda}\cdot \sin(\lambda L)\cdot\left[\eta(z-H)\cdot \mathrm{sh}(\lambda(z-H)) - e^{-\lambda H}\mathrm{sh}(\lambda z)\right]. \tag{2.10}$$

Let us sum (2.9) and (2.10) and fulfill the transverse Fourier transform according to formula (2.5). The result must be substituted into Maxwell Equations (2.2) and (2.3).

After the transfer from L—image to the space of originals, we will obtain the expressions for the magnetic field intensity components excited by the linear currents in the space between the orthogonal ideal conducting planes $(z=0\perp y=0, z>0,\ y>0)$:

$$\begin{aligned} H_y(t,y,z) = -\frac{2}{\pi}I(t)\int\limits_0^{\infty} &\{\sin(\lambda\ell)\cdot\left[\eta(z-h)\cdot \mathrm{ch}(\lambda(z-h)) - e^{-\lambda h}\mathrm{ch}(\lambda z)\right] \\ &- \sin(\lambda L)\cdot\left[\eta(z-H)\cdot \mathrm{ch}(\lambda(z-H)) - e^{-\lambda H}\mathrm{ch}(\lambda z)\right]\}\cdot \sin(\lambda y)\cdot d\lambda \end{aligned} \tag{2.11}$$

$$\begin{aligned} H_z(t,y,z) = -\frac{2}{\pi}I(t)\int\limits_0^{\infty} &\{\sin(\lambda\ell)\cdot\left[\eta(z-h)\cdot \mathrm{sh}(\lambda(z-h)) - e^{-\lambda h}\mathrm{sh}(\lambda z)\right] \\ &- \sin(\lambda L)\cdot\left[\eta(z-H)\cdot \mathrm{sh}(\lambda(z-H)) - e^{-\lambda H}\mathrm{sh}(\lambda z)\right]\}\cdot \cos(\lambda y)\cdot d\lambda \end{aligned} \tag{2.12}$$

The tangent components of intensity on the metal surfaces is found from dependencies (2.11) and (2.12) by substituting $z=0$ and $y=0$.

After introducing a new variable of integration $x = \lambda \cdot h$ and fulfilling necessary transformations, expressions may be obtained that are comfortable for the practical calculations.

$$H_y(t, y, z = 0) = H_m(t) \cdot \int_0^{\infty} \left(\mathrm{e}^{-x} \sin\left(x \cdot \frac{\ell}{h} \right) - \mathrm{e}^{-x \cdot \frac{H}{h}} \sin\left(x \cdot \frac{L}{h} \right) \right) \cdot \sin\left(x \cdot \frac{y}{h} \right) \cdot \mathrm{d}x, \tag{2.13}$$

where $H_m(t) = \frac{2 \cdot I(t)}{(\pi \cdot h)}$.

$$H_z(t, y = 0, z) = -H_m(t) \int_0^{\infty} \left\{ \left(\eta(z - h)\mathrm{sh}\left(x\left(\frac{z - h}{h} \right) \right) - \mathrm{e}^{-x}\mathrm{sh}\left(x\frac{z}{h} \right) \right) \sin\left(x\frac{\ell}{h} \right) - \left(\eta(z - H)\mathrm{sh}\left(x\left(\frac{z - H}{h} \right) \right) - \mathrm{e}^{-x}\frac{H}{h}\mathrm{sh}\left(x\frac{z}{h} \right) \right) \sin\left(x\frac{L}{h} \right) \right\} \cdot \mathrm{d}x. \tag{2.14}$$

Expressions (2.13) and (2.14) may be integrated analytically. This is the first integration that simplifies their view.

As follows from the practice of magnetic pulsed metal-working (EMF), the chosen gap values are the same between the inductor's first turn, which is placed in the direct nearness from the angle and by the orthogonal planes. Thus, a non-destruction generality of the consideration being conducted, $\ell = h$, may be suggested. The coordinates of the inductor's second turn may be described by the dependences according to Fig. 2.1: $L = h \cdot \left(1 + \frac{D}{h} \cdot \cos \alpha\right)$, $H = h \cdot \left(1 + \frac{D}{h} \cdot \sin \alpha\right)$.

After all above-mentioned operations, the spatial distribution of magnetic fields on the metal surfaces of the angle profile will be as follows:

$$H_y^0(y, z = 0) = \left(\frac{y}{h} \right) \cdot \left[\frac{1}{\left(2 + \left(\frac{y}{h} \right)^2 \right)^2 - 4\left(\frac{y}{h} \right)^2} - \frac{\left(1 + \frac{D}{h}\sin\alpha \right)\left(1 + \frac{D}{h}\cos\alpha \right)}{\left(\left(\frac{y}{h} \right)^2 + \left(1 + \frac{D}{h}\sin\alpha \right)^2 + \left(1 + \frac{D}{h}\cos\alpha \right)^2 \right)^2 - 4\left(\left(\frac{y}{h} \right)^2 \left(1 + \frac{D}{h}\cos\alpha \right)^2 \right)} \right], \tag{2.15}$$

where $H_y^0(y, z = 0) = \frac{H_y(t, y, z=0)}{\left(\frac{I(t)}{(\pi \cdot h)} \right)}$.

$$H_z^0(y=0,z) = -\left(\frac{z}{h}\right)\cdot\left[\frac{1}{\left(2+\left(\frac{z}{h}\right)^2\right)^2-4\left(\frac{z}{h}\right)^2}\right.$$
$$\left.-\frac{\left(1+\frac{D}{h}\sin\alpha\right)\left(1+\frac{D}{h}\cos\alpha\right)}{\left(\left(\frac{z}{h}\right)^2+\left(1+\frac{D}{h}\sin\alpha\right)^2+\left(1+\frac{D}{h}\cos\alpha\right)^2\right)^2-4\left(\left(\frac{z}{h}\right)^2\left(1+\frac{D}{h}\sin\alpha\right)^2\right)}\right], \tag{2.16}$$

where $H_z^0(y=0,z) = \frac{H_z(t,y=0,z)}{\left(\frac{I(t)}{(\pi\cdot h)}\right)}$.

Truthfulness of the expressions found for the magnetic-field components can be seen in their symmetry relative to the orthogonal surfaces $z=0$ and $y=0$.

If the entire system is turned around on an angle of 90° in the positive direction, the horizontal plane will accept the position of the vertical one. The vertical plane will become the horizontal one. The trigonometric-function argument is changed on the quantity $\sim\frac{\pi}{2}$. Thus, exactness of formula (2.15) passes in the expression (2.16) and, vice versa, the formula (2.16) in formula (2.15).

The sign change in front of the expressions for the intensity components is caused by change of the exciting current orientation relative to the planes forming the right angle.

In addition, formulas (2.15) and (2.16) coincide with each other if $\alpha = 45°$ (again, with exactness to their sign, which is fixed by the direction taken by the current in the current-conductor). This fact means that there is "equality of rights" of the boundaries of the problem being considered.

This inference also follows from the simple physical considerations: the excitation processes of the whirlwind currents must be identical in the system, which is geometrically symmetrical relative to the angle bisector.

Thus, under the skin-effect condition, the expressions for the magnetic-pressure forces on the conducting planes $z=0$ and $y=0$ are written as follows:

(a) the horizontal plane,

$$P_z^0(y,z=0) = \left(H_y^0(y,z=0)\right)^2, \tag{2.17}$$

where $P_z^0(y,z=0)$—is the relative pressure,

$$P_z^0(y,z=0) = \frac{\left(\frac{\mu_0 H_y^2(t,y,z=0)}{2}\right)}{\left(\frac{\mu_0 H_m^2(t)}{2}\right)},\ H_m(t) = \frac{I(t)}{(\pi\cdot h)}$$

(b) the vertical plane,

$$P_y^0(y = 0, z) = \left(H_z^0(y = 0, z)\right)^2, \tag{2.18}$$

where $P_y^0(y = 0, z)-$ is the relative pressure,

$$P_y^0(y = 0, z) = \frac{\left(\frac{\mu_0 H_z^2(t, y=0, z)}{2}\right)}{\left(\frac{\mu_0 H_m^2(t)}{2}\right)}.$$

Formulas (2.15) through (2.18) fully describe the distributions of the magnetic field–intensity vector and the pressure distributions on the surfaces of the sheet-metal work-piece angle zone.

2.3 Numerical Illustrations and Some Conclusions

To determine the quantity laws of the electrical dynamic processes being investigated, a concrete example from magnetic pulsed metal processing is considered.

Let $h = 0.0005$ m, $D = 0.02$ m, $\alpha =$ var, and $\alpha \in [0, 15°]$.

A graphical illustration of the results calculated is represented in Figs. 2.2 and 2.5.

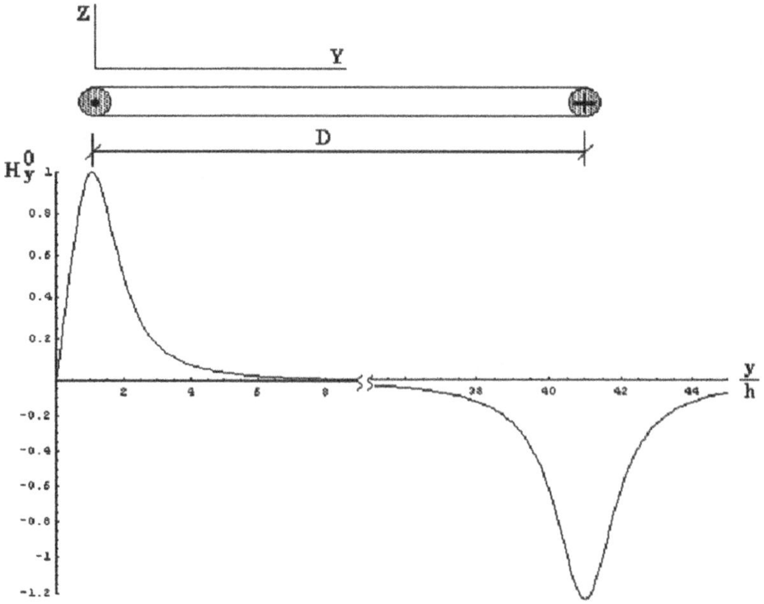

Fig. 2.2 Magnetic-intensity tangent-component distribution along the horizontal surface calculated by formula (2.15) for $\alpha = 0$

The first components to be marked on the graphical drawings (for the most visibility!) are the intensity values normalized on the value of its maximum calculation for the tangent component of the field excited by the current-conductor $(h,\ h)$ on the plane $z = 0$. That is, the dependencies for $H^0_{y,z} = \frac{H_{y,z}}{H_{m1}}$ are represented on the graphics. In the present calculations we have $H_{m1} \approx 0.2 \cdot H_m(t) = 0.2 \cdot \frac{I(t)}{(\pi h)}$. For the current amplitude ≈ 100 kA and $h = 0.0005$ m, we have $H_{m1} \approx 1.27 \cdot 10^7 \frac{\mathrm{A}}{\mathrm{m}}$. The according value of the pressure will be equal to $P_{m1} \approx 101.86\,\mathrm{MPa}$.

A second important remark also should be noted: The intensity calculated on the vertical plane for $\alpha = 0$ by the adopted formulation of the problem does not have a strict physical sense.

As is known for flowing current, a closed circuit is necessary, which must reflect the geometrical shape of the initial exciting signal. The adopted calculation model (see Fig. 2.1) suggests that current wires have an infinite length in the longitudinal direction. Physically this means that somehow they form somewhere some closed circuits for flowing the induced currents ways on planes $z = 0$ and $y = 0$.

From a mathematical point of view, it is not important whether these current wires are part of the single circuit or parts of the separate turns that lie (or do not lie) in the same plane.

The problem with the adopted mathematical formulation accords strictly to the electromagnetic interaction only of the parallel conductors with opposite directed currents placed between the orthogonal ideally conducting planes.

Some physical interpretation of the obtained solutions must be made for states that are naturally adequate to the mathematical description of the flowing processes. Thus, already for the angle of the inductor turn slope $\alpha = 15°$, the field calculation on the vertical plane is adequate to the processes in the inductor system being considered. It reflects the electromagnetic connection between the exciting initial signal and the induced currents on both the horizontal and vertical planes. The geometry for $\alpha \neq 0$ will accord to the physical state when the parallel current-conductors form one closed circuit placed according to angle of the slope. The according closed circuit for the flowing of whirlwind currents appears on the vertical plane. This circuit represents a geometrical reflection of the exciting signal.

To illustrate the above-mentioned considerations of an adequate picture of the magnetic field–intensity tangent component along with the vertical for the angle zero slope of the turn plane to the horizontal surface of the sheet work-piece is given in Fig. 2.6.

On the whole, the calculations lead to the following inferences.

- The current of the parallel thin current wires of the rectangular turn, which was placed in the inner cavity of the plane angle of the bent metal sheet with the high specific electrical conductivity, excites on the coplanar planes an intensive magnetic fields and great pressure forces with the same time shape. The intensity $\sim 10^7 \frac{\text{A}}{\text{m}}$ for the current character value $\sim 100\,\text{kA}$ works in the magnetic pulsed metal. The pressure maximums (up to $\sim 100\,\text{MPa}$) of the space distribution will be localized in points with coordinates according to arrangements of the current-conductors (Figs. 2.2 and 2.3).

- The Figs. 2.2 and 2.3 exceeding the value of the field and pressure being excited in the “farther” current wire zone (L, H) above the according value in the “nearer” current wire zone (ℓ, h) may have a place on practice really. One should be taken into account that a truly strict coplanar wire placement is impossible. Some distortion in the relative arrangement of the inductor turn and planes will always occur. In this case, the exceedance may be explained by the dissipation of the current energy of the “closer” current wire for field excitation—as in the quasi-coplanar plane metal—on the orthogonal surface of the angle of the bent work-piece.
- Increasing the slope angle of the turn plane to the according surface up to approximately 15° leads to essentially decreasing the excited field intensity amplitude (approximately by a factor of 10) as well as the magnetic pressure value ($\sim$by a factor of 100) of the “farther” current-conductor zone (Figs. 2.3 and 2.4). Physically, the given fact can be explained by an increase of the electromagnetic connection between the current and the conductor because of the distance increasing between them (the confirmation is seen in Figs. 2.5 and 2.6).

Variation of the inter-arrangement of the turn and the one of the plane angle surfaces of the bent sheet work-piece allows the realization of the essential concentration of the field and pressure in the immediate nearness of the bend.

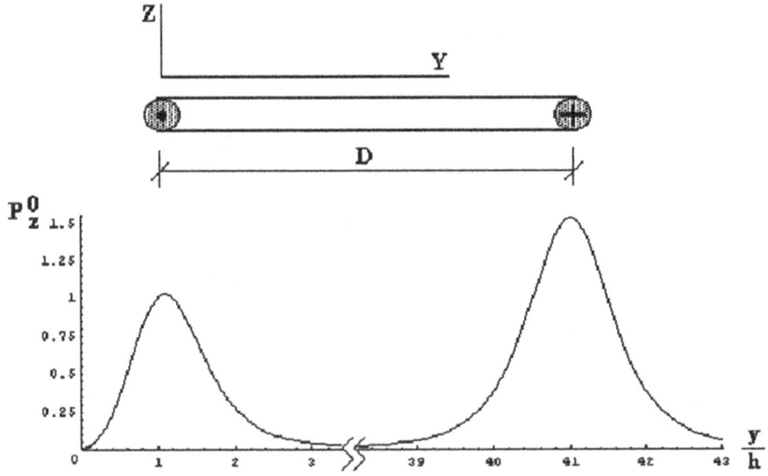

Fig. 2.3 Distribution of the magnetic-pressure normal component along the horizontal surface calculated by formula (2.17) for $\alpha = 0$

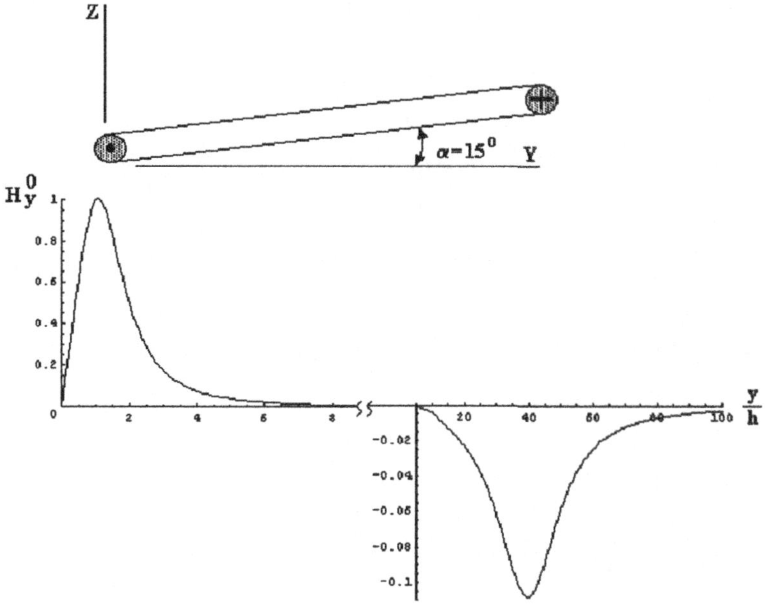

Fig. 2.4 Distribution of the magnetic-intensity tangent component along the horizontal surface calculated by formula (2.15) for $\alpha = 15°$

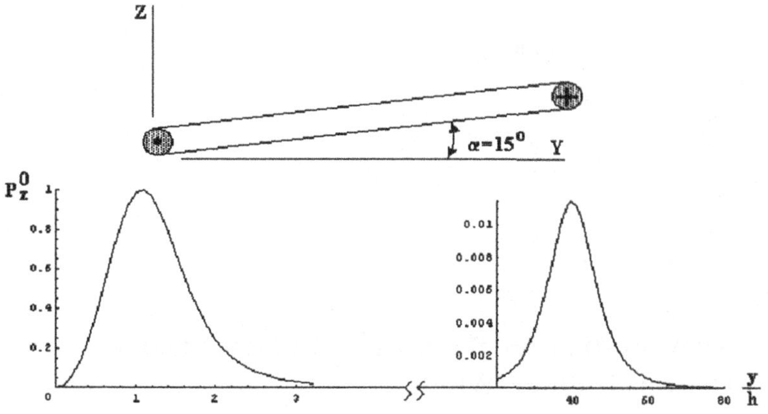

Fig. 2.5 Distribution of the magnetic-pressure normal component along the horizontal surface calculated by formula (2.17) for $\alpha = 15°$

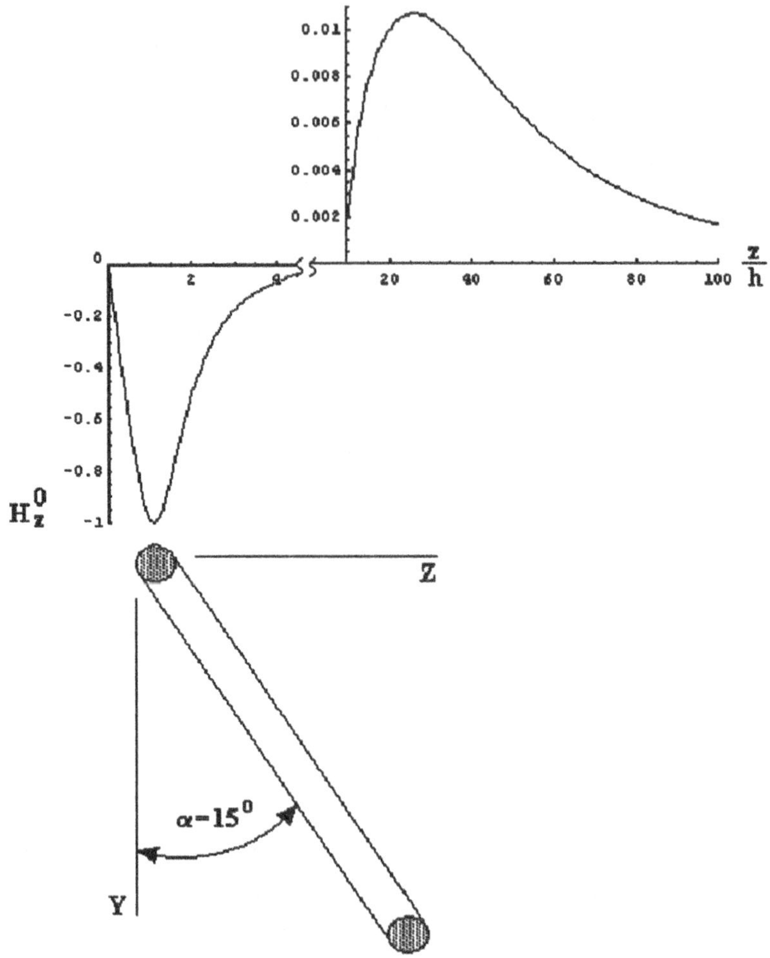

Fig. 2.6 Magnetic-intensity tangent component distribution along the vertical surface as calculated by formula (2.16) for $\alpha = 15°$

2.4 Generalization for the Case of Inductor-Current Wires with Finite Width

In the previous case, the current wires were quite thin in the cross-section. However, the cross-dimensions of the field sources have an influence on the excitation processes of the inductor system. Obviously, the field and pressure amplitudes will have to be different, and the distribution shapes will have to be different as well, etc. All of this demands additional evaluations.

Figure 2.7 shows the calculation model in which the current wires have some specified finite width with negligible thickness. Take into account that the last geometrical parameter is not needed. Because of the electrical magnetic symmetry in relation to the angle between the conducting surfaces, this model of the interacting work-piece with a horizontal plane provides answers for the questions formulated.

The magnetic field–intensity tangent component on the plane $z = 0$ can be found from expression (2.13) where the Fourier-transformations of the y-distribution of the exciting current according to the model in Fig. 2.7 should be introduced [the function in formula (2.6) is $f_{1,2}(\lambda)$].

Finally, the given dependence for the practical calculation by $\ell = h$ will appear as follows:

$$H_y^0(y, z=0) = \int_0^\infty \frac{e^{-x}}{x} \cdot \left[\sin\left(y\left(1+\frac{a}{2h}\right)\right) \cdot \sin\left(y \cdot \frac{a}{2h}\right) - \left(\frac{a}{b}\right) e^{-y\left(\frac{D}{h}\right)\sin\alpha} \sin\left(y\left(1+\frac{a+b}{2h}+\frac{D}{h}\cos\alpha\right)\right) \cdot \sin\left(y \cdot \frac{b}{2h}\right) \right] dx, \tag{2.19}$$

where $H_y^0(y,\ z=0) = \frac{H_y(t,\ y,\ z=0)}{\left(\frac{I(t)}{\pi \cdot a}\right)}$.

Naturally, if $a \to 0$ and $b \to 0$, the expression (2.19) passes expression (2.13).

The magnetic pressure on the horizontal surface of the angle can be calculated with help of the formula (2.17).

Figures 2.8 and 2.9 show graphic illustrations of the calculation results of the intensity and the pressure space distribution for an inductor system with the following geometrical parameters: $a = 0.005$ m, $D = 0.02$ m, $b =$ var, $\alpha =$ var.

Similar to the previous calculations, the intensities were normalized on the value of the maximum of the field tangent component of the "near" current-conductor.

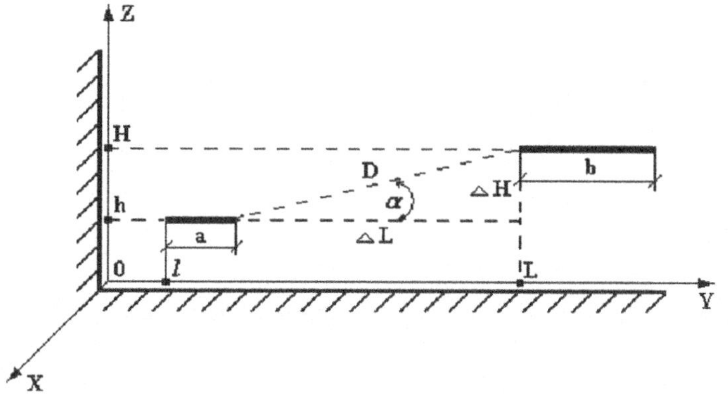

Fig. 2.7 Cross-section of an inductor system with a single-turn rectangular solenoid, the current-conductors of which have finite wideness, in an inner cavity of the angle in a curved metal sheet

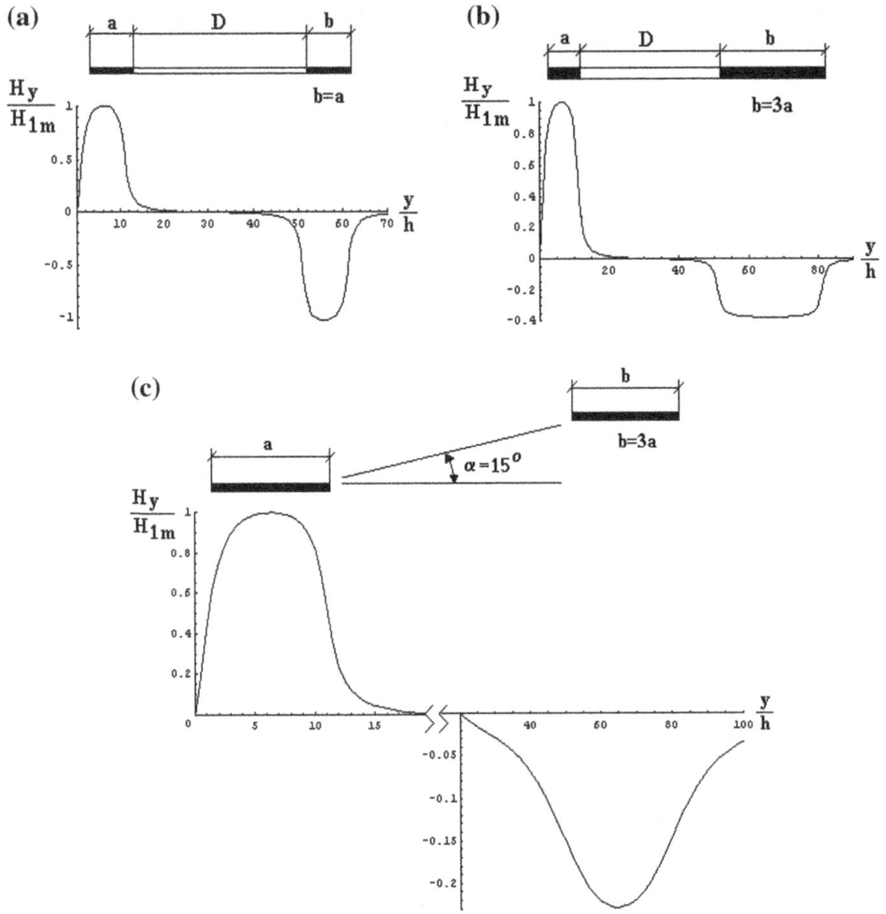

Fig. 2.8 Magnetic intensity tangential component distribution along the horizontal surface as calculated by formula (2.19): **a** $\alpha = 0, b = h$; **b** $\alpha = 0, b = 3 \cdot h$; and **c** $\alpha = 15°, b = 3 \cdot h$

The dependences for $H^0_{y,z} = \frac{H_{y,z}}{H_{m1}}$ are represented on the graphics. In the present calculations, the normalization equals $H_{m1} \approx 0.665 \cdot \frac{I(t)}{(\pi \cdot a)}$. For a current value of $\approx$100 kA, we have $H_{m1} \approx 0.423 \times 10^7 \frac{\text{A}}{\text{m}}$. The according value of pressure will equal $P_{m1} \approx 11.3\,\text{MPa}$.

Inferences from calculations are as follows.

- Fulfilling the current wire with finite width essentially decreases the amplitudes of the fields excited and their pressures. Thus, on passing from the point to strip model in the calculations, the maximum decrease of the magnetic field intensity was fixed at a maximum decrease of $\sim$33% and that of the pressure at $\sim$89%. The given fact can be easily explained: The point model and, consequently, the

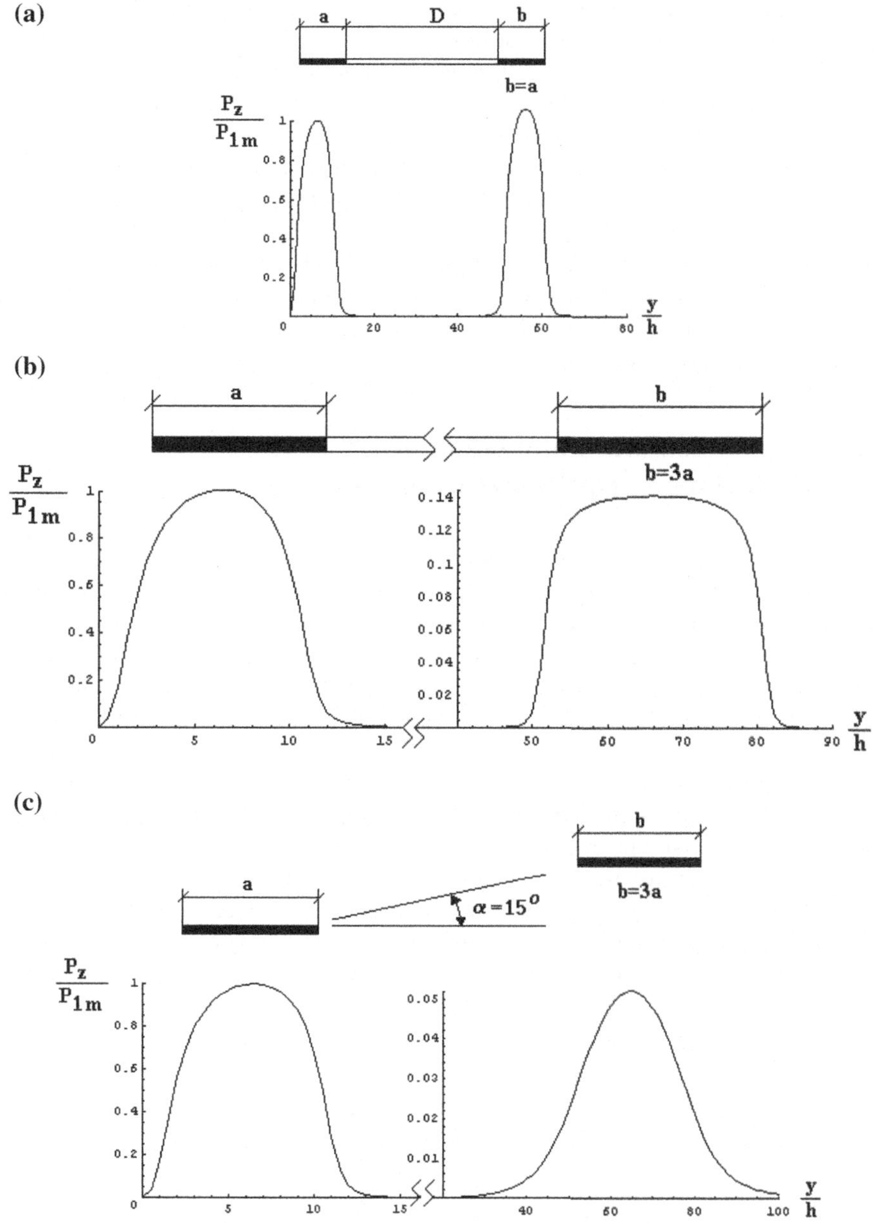

Fig. 2.9 Magnetic pressure normal component distribution along the horizontal surface as calculated by formulas (2.19) and (2.17): **a** $\alpha = 0, b = h$; **b** $\alpha = 0, b = 3 \cdot h$; and **c** $\alpha = 15°, b = 3 \cdot h$

greatest intensities occur as the correlation of the width and the distance to the conducting plane approaches zero.

- In the case of current-wires with finite width in excess of the second maximum above the first one, intensity as well as pressure will essentially be lower than in the case of thin conductors (Figs. 2.8a and 2.9a).
- Increasing the width of the "farthest" current-wire, compared with the "closer" one, leads to an essential decrease of field intensity and pressure forces from its side (Figs. 2.8b and 2.9b).
- A combination of the plane-slope angle, where the current-wires are situated, and the width of the "farthest" current wire permits realizing the concentration of quite a high level of the field and pressure in the angle zone of the bent metal sheet (Figs. 2.8c and 2.9c).

2.5 Suggestion and Foundation of Effective Inductor-System Construction

A drawback of the considered inductor-system option that is destined for the angles "filling in" in the bent metal sheet is that the forces are concentrated near the angle but not in its centre. They direct not to the centre but to the planes limiting the angle. However, theoretically it can be expected that an equal-acting force of magnetic pressure created by the sloped turn of the inductor on the orthogonal planes in the "closer" current-conductor area will be directed toward the centre.

Concentrating and directing the forces to the centre may be achieved if another linear conductor, placed symmetrically to the existing "farther" current wire relative to the angle bisector of the curved metal sheet, is introduced in the inductor scheme in Fig. 2.1. The current in the introduced current wire must be the same as the current in the current wire of the "closer" zone but oppositely directed.

Schematically, the proposed inductor system is represented in Fig. 2.10.

After fulfilling all necessary mathematical transformation in expressions (2.15) and (2.16), we obtain the intensities of the field that was excited in the inductor system shown in Fig. 2.10:

$$H_y^0(y, z=0) = \left(\frac{y}{h}\right) \cdot \left[\frac{1}{\left(2+\left(\frac{y}{h}\right)^2\right)^2 - 4\left(\frac{y}{h}\right)^2} - \frac{\left(1+\frac{D}{h}\sin\alpha\right)\left(1+\frac{D}{h}\cos\alpha\right)}{\left(\left(\frac{y}{h}\right)^2 + \left(1+\frac{D}{h}\sin\alpha\right)^2 + \left(1+\frac{D}{h}\cos\alpha\right)^2\right)^2 - 4\left(\left(\frac{y}{h}\right)^2\left(1+\frac{D}{h}\cos\alpha\right)^2\right)} - \frac{\left(1+\frac{D}{h}\sin\alpha\right)\left(1+\frac{D}{h}\cos\alpha\right)}{\left(\left(\frac{y}{h}\right)^2 + \left(1+\frac{D}{h}\sin\alpha\right)^2 + \left(1+\frac{D}{h}\cos\alpha\right)^2\right)^2 - 4\left(\left(\frac{y}{h}\right)^2\left(1+\frac{D}{h}\sin\alpha\right)^2\right)} \right]. \tag{2.20}$$

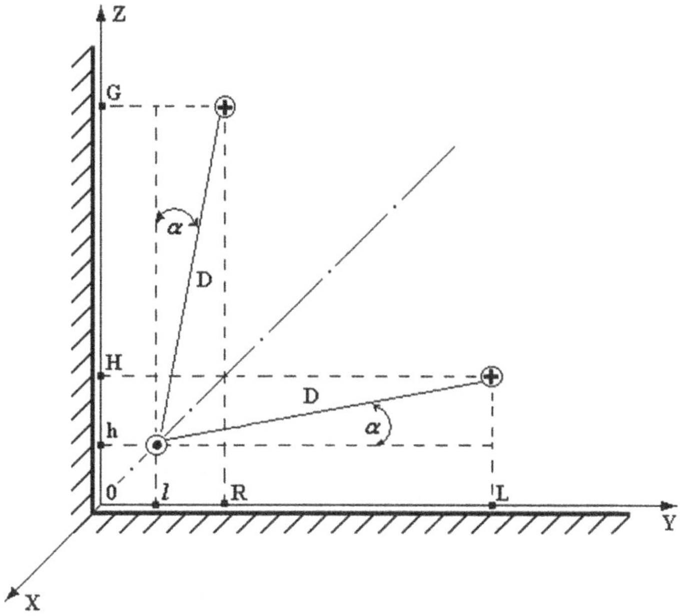

Fig. 2.10 Cross-section of a schematic model of an inductor system with two single-turn rectangular solenoids with common sides in an inner zone of the angle of a flat metal sheet

$$H_z^0(y=0,z) = -\left(\frac{z}{h}\right)\cdot\left[\frac{1}{\left(2+\left(\frac{z}{h}\right)^2\right)^2-4\left(\frac{z}{h}\right)^2} - \frac{\left(1+\frac{D}{h}\sin\alpha\right)\left(1+\frac{D}{h}\cos\alpha\right)}{\left(\left(\frac{z}{h}\right)^2+\left(1+\frac{D}{h}\sin\alpha\right)^2+\left(1+\frac{D}{h}\cos\alpha\right)^2\right)^2-4\left(\left(\frac{z}{h}\right)^2\left(1+\frac{D}{h}\sin\alpha\right)^2\right)} - \frac{\left(1+\frac{D}{h}\sin\alpha\right)\left(1+\frac{D}{h}\cos\alpha\right)}{\left(\left(\frac{z}{h}\right)^2+\left(1+\frac{D}{h}\sin\alpha\right)^2+\left(1+\frac{D}{h}\cos\alpha\right)^2\right)^2-4\left(\left(\frac{z}{h}\right)^2\left(1+\frac{D}{h}\cos\alpha\right)^2\right)}\right]. \tag{2.21}$$

The pressure forces can be calculated by formulas (2.17) and (2.18) with the help of formulas (2.20) and (2.21).

As follows from the calculation results in the suggested inductor system, represented graphically in Fig. 2.11, the same magnetic forces will act on the surface near the angle bend of the metal sheet. Their equal-acting force will be strictly directed to the angle centre, and its amplitude exceeds the times of the coordinate component amplitude.

The drawback to this is a relative complication of the inductor-making caused in the first turn by the requirement of nearness of the parallel main current wires and by the low electrical strength of this construction. However, in this case the advantage consists of the excited fields and pressures with quite high amplitudes. In accordance with the circuit shown in Fig. 2.12b, the same currents flow in each turn. In the first approximation, it can be suggested that each turn works independently of the other and induces according currents and fields in the sheet metal. In the zone of the main current wire, their amplitudes will be defined by superposition of the currents and the fields the being generated by each turn separately.

The intensity amplitude increases by 2 times, but the pressure by 4 times, compared with parameters of the processes with the current $I\ (t)$ being in the main current-conductor. Thus, taking in account the finite width of the main current wire for the current ~ 100 kA, the intensity of the magnetic field will be reached —$H_{m1} \approx 0.846 \times 10^7\ \frac{\text{A}}{\text{m}}$. The maximum magnetic pressure equal-acting force will equal ~ 63.81 MPa.

The main results of the conducted consideration can be united by the following positions.

1. Theoretical analysis is performed of the electrical-dynamics processes in the inductor systems' different options destined for "filling in" of the angles in the bent sheet metal.
2. For the optimum option of the inductor system with two separate turns of the rectangular geometry that are inputted serially to the power source (the magnetic pulsed installation), the excited force will be directed to the centre of the bend of the metal sheet. The amplitude of this force can reach tens of MPa.

2.6 Experimental Investigations of Inductor Systems for Stamping Angles in Sheet Metal

This section is devoted to the description of the experiments with inductor systems in which an aim-directed force action of rather high electrical conductivity occurs on the inner cavity of the flat angle in the curved sheet metal.

The equipment used includes the following:

- A generator of harmonic current with a working frequency of ~ 5.0 to 40 kHz (Power Ron 80S, Industrial Test Equipment Co., Port Washington, NY).
- An oscillograph (four-channel, digital-storage oscilloscope [Tektronix TDS 2014])
- Induction probes (dimensions of 0.002×0.002 m^2) to measure the currents and the magnetic-field intensities.
- A model of the tool: a longitudinal lengthy single-turn solenoid made from cooper wire of diameter ∅0.0015 m with the dimensions 0.15×0.02 m^2.
- An aluminum sheet with the bend angle near $\sim \frac{\pi}{2}$ and thickness of ~ 0.001 m. The dimensions of the angle planes were the same and equaled 0.2×0.06 m^2.

The scheme of the first experiment is shown in Fig. 2.13.

The sine current with frequency $f = 20.6$ kHz was given to the input of the inductor. The signal from the induction probes was given to the input of the oscillograph.

The concrete aim of the experiments was measuring of the current density and of the magnetic-field intensity on the surfaces of the sheet billet. The first group of experiments was dedicated to measuring the characteristics of the processes in the case when the inductor plane is strictly coplanar to the horizontal plane of the angle. The measurement results are shown in Fig. 2.14.

The measurement results are as follows:

- The density of the excited current has two local maximums on the horizontal plane of the bend. The first of them is situated close to the angle under the "nearest" current wire (position № 2 in Fig. 2.14a). The second one is situated under the "farthest" current wire of the rectangular turn (position no. 9 in Fig. 2.14a).
- The value of the second maximum exceeds the first one by approximately 1.25 times, which can be explained by dissipation of the electromagnetic energy from the "nearest" turn on exciting the field in metal of the bent orthogonal surface.
- The maximum of the excited field on the vertical plane (position no. 2 in Fig. 2.14b) is 70% of the analogical parameter for the field of the "nearest" turn.

The conditions of the experiment being conducted accords practically to the skin-effect regime. That is why the estimates of the electrodynamical efforts can be

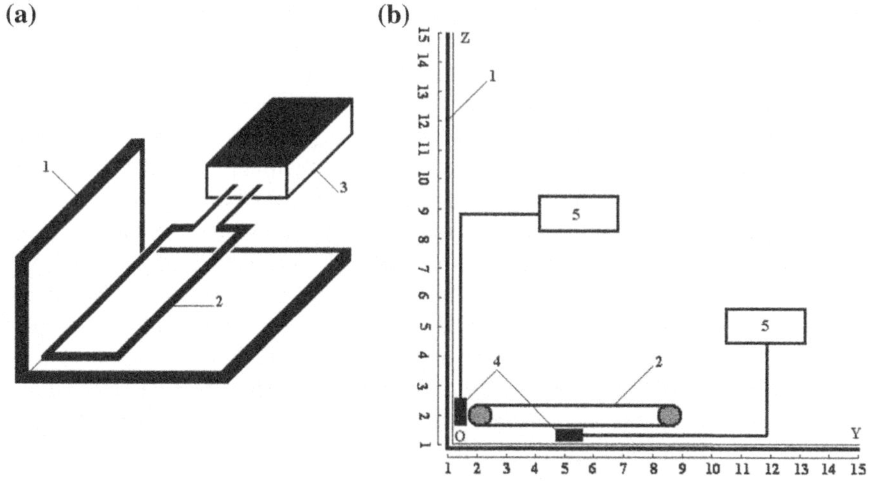

Fig. 2.13 The experimental system (**a**) scheme of the experiment (**b**) cross-section of the experimental system in the zone of measuring. 1 the bent sheet billet; 2 the inductor; 3 the generator of the harmonic current; 4 the induction coil probes; and 5 the oscillograph

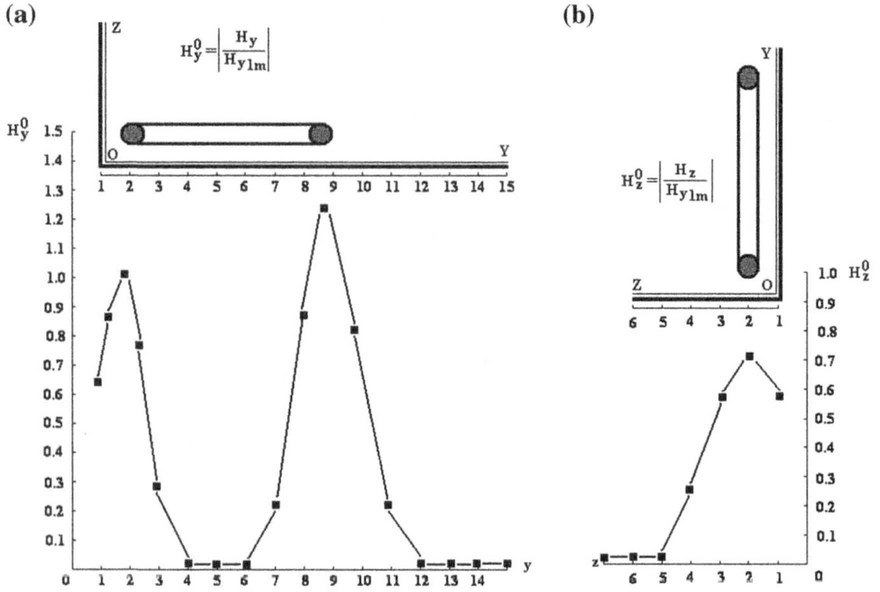

Fig. 2.14 The space distribution of the relative module of the magnetic intensity tangent component on the angle surfaces of the bent sheet (the maximum value of the intensity was accepted as the norm) (**a**) the horizontal surface (**b**) the vertical surface

fulfilled by considering that the pressure-forces amplitude will be proportional to the squares of the according components of the magnetic-field intensities.

Thus, in the angle zone the ratio of the pressure forces acting on the horizontal and vertical planes will approach ~2. Their resultant point of intensity maximums (crossing of straight lines by position no. 2 on both Fig. 2.14a, b) will be directed not in the side of angle but with the essential displacement along the horizontal plane.

On the metal of the horizontal surface, the pressure acts with amplitude exceeding the maximum in the angle zone by more than approximately 1.56 times (position no. 2 in Fig. 2.14a). In general, the essential drawback of the investigated construction of the inductor system is excitation of the pressure forces acting on the sheet billet outside of the zone where formation of a clear angle is required.

This drawback can be ameliorated (by conserving one solenoid and decreasing the pressure-forces amplitude outside the predetermined area) if the turn plane is situated under some angle to the horizontal surface of the bend.

The second group of experiments was performed for the case when the inductor plane is situated to the bend surface under an angle of approximately 15°. The results are shown in Fig. 2.15.

As follows from the physical considerations, the magnetic-field amplitude on the metal surface must decrease in order to increase the distance from the current-wire to the metal. This fact was being confirmed in the experiments. In the zone under the "farthest" current-wire (position no. 9 in Fig. 2.15a), the intensity value

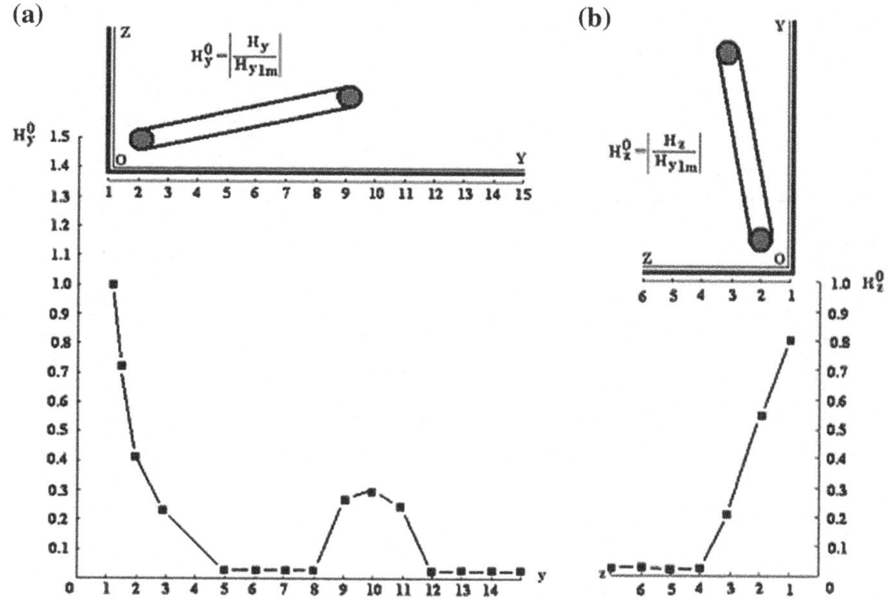

Fig. 2.15 Relative module of the magnetic intensity tangent component when the inductor plane is situated to the bend surface under an angle of ~15° (the maximum value of the intensity was accepted as the norm) (**a**) the horizontal surface (**b**) the vertical surface. Note that a ratio of the field intensities equals ~0.84 for the bend angles of ~15° and ~0°

decreases by more than approximately 4 times. Accordingly, the pressure force decreases by more than 16 times. In addition, some displacement of the maximum points occurs in the direction of the angle centre.

The amplitude of the tangent component of the magnetic field vector on the vertical plane of the bent metal increases by almost ~1.15 times (Fig. 2.15b).

The measurements initiated additional experiments with further increasing of the inductor plane–slope angle. Naturally, the increase is limited by a value of ~45°. Further increase leads to a switch in places of the horizontal and vertical planes of the bend during the excitation process of eddy currents in the metal sheet.

Figure 2.16 shows the space distribution of the magnetic field–intensity tangent components on the angle-bend surfaces and, accordingly, the induced current-density distribution under increasing inductor plane slope angle up to ~45°.

In the present constructive fulfillment of the inductor system, it can be expected that the resultant pressure forces acting on the orthogonal surfaces of the bend will be directed strictly to its centre. That is why practical aim-directed magnetic pressure to the angle is possible. As a result, its clear formation is possible although the amplitude of excited forces decreases by approximately 2 times.

A main drawback of this single-turn inductor system with the slope solenoid is the rather weak electrodynamic tie of exciting current in the turn and the current induced in the metal sheet. Consequently, this results in the aforementioned

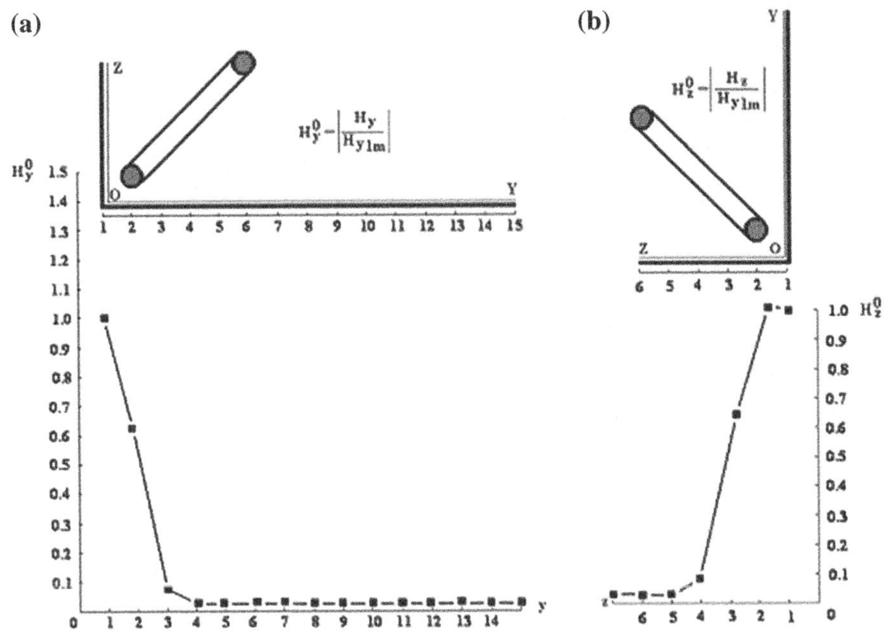

Fig. 2.16 The module of the relative value of the magnetic field intensity tangent component under the inductor plane–slope angle up to ~45° to the horizontal plane (the maximum value of the intensity was accepted as the norm) (**a**) the horizontal surface (**b**) the vertical surface. Note that a ratio of the field intensities equals ~0.6 for bend angles of ~45° and ~0°

essential decreases of eddy-current amplitude and pressure forces. In addition, precise adjustment of the solenoid provides magnetic pressure strictly to the centre, which is a very complicated but practical task.

The inductor system with so-called double-slope solenoids is steadier and more effective in relation to the magnetic field–source position for the achievement of rather high amplitude of the forces being excited (Fig. 2.17a).

Constructively, such a system is fulfilled by the formation of two lengthy rectangular turns located under a predetermined angle to each other. The turns are connected mechanically along the one long side. There can be electrical contact between them or not. In accordance with the connection method (between the turns) as well as their attachment to the power source, the turns can form parallel or consequent circuits.

In the first case, the source current will flow along the central current-wire uniting two long branches of the turns. If the lengthy branches are parallel and divided, the flowing current will be less by 2 times. Accordingly, the magnetic-pressure forces outside the angle zone being formed decrease by almost 4 times.

Under consequent connection of the turns, the current value in the central current-wire doubles. The pulse current will flow by the lengthy parallel branches, and its amplitude will be equal to that of the source current. In general, effectiveness

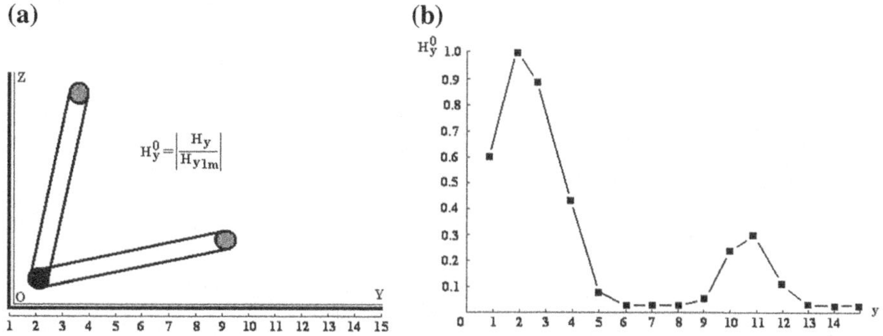

Fig. 2.17 Inductor system (**a**) and module of the normed amplitude of the magnetic field intensity tangent component on the angle surfaces in the bend of the metal sheet (**b**) (the maximum value of the intensity was accepted as the norm)

under such a connection of turns increases by more than 4 times compared with the option of their connection mentioned previously. It should be noted that an analogy was suggested previously for the copper-foil stamping (the separate rectangular turns having one common side). Such tools have been termed "inductor systems with doubled current-wires."

The results of the magnetic field intensity measure in the space between the inductor current-wires and the surfaces of the sheet billet are represented in Fig. 2.17b.

The turns were connected consecutively. Their angle of slope to the according planes was strictly equal to ~15°, and their arrangement was symmetrical relative to the horizontal and vertical planes.

The previous experiments were conducted for sheet billets with flat, straight angles. In fact, the straight angle must be formed with help of magnetic pressure on the bends obtained by rounding or other production methods. In fact, this is real billet because the consequent magnetic pulsed processing is the metal sheet, in the bend of which some part of a cylindrical surface exists.

In this connection, the question about the influence of rounding on the characteristics of the electromagnetic processes represents a practical interest. The consequent experiments were conducted with a billet, the perpendicular planes of which in the bend zone were joint with rounding along a radius of approximately 0.01 m.

As it follows from measures in the investigated inductor system—as in the case of small rounding (Fig. 2.17b), so it is in the case of large rounding (Fig. 2.18)—the main results can be united by the next positions.

- The resultant force of the magnetic pressure in the working zone is directed strictly to the centre of the angle being formed, and its amplitude will exceed by more than 1 order of the analogical characteristic that of the force action outside the working zone.

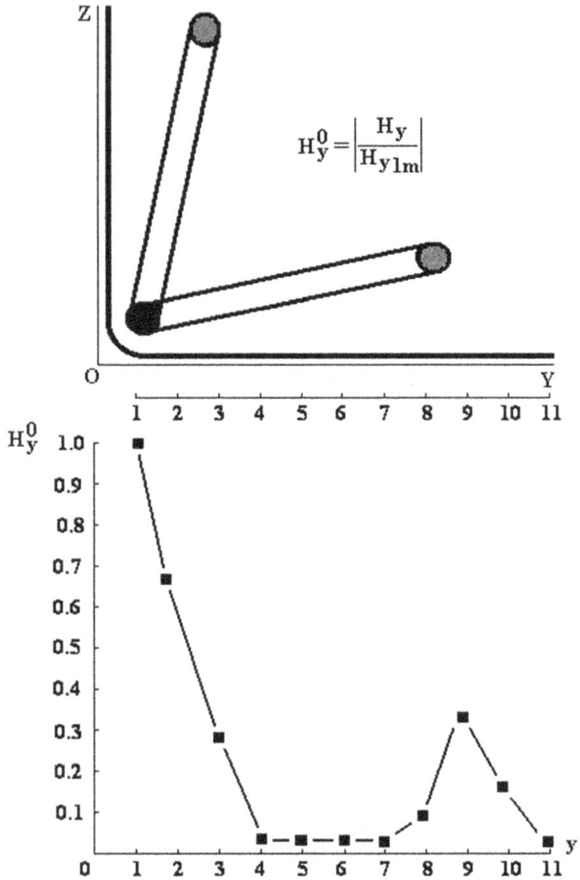

Fig. 2.18 Module of the normed amplitude of the magnetic field intensity tangent component on the angle surfaces in the bend with rounding along a radius of approximately 0.01 m (the maximum value of the intensity was accepted as the norm)

- Under consequent connection of the inductor-system turns, the effectiveness of the force action increases essentially at the expense of the current doubling in the central current-wire located along the bend line.

In the conducted experiments, the tangent component of the magnetic field intensity was measured in a space between the inductor turns and the sheet billet. Its value was interpreted in term of the induced current. However, the measured intensity is a field characteristic in the air gap but not strictly on the metal surface. That is why the direct measures of current induced in the billet metal were performed for confirmation of the conclusion's reliability about the investigated electromagnetic processes. Visually the conducted experiment is shown in Fig. 2.19.

All options have been investigated of inductor systems with one rectangular single turn located under different angles to the horizontal plane of the sheet billet bend and with two rectangular turns consequently connected.

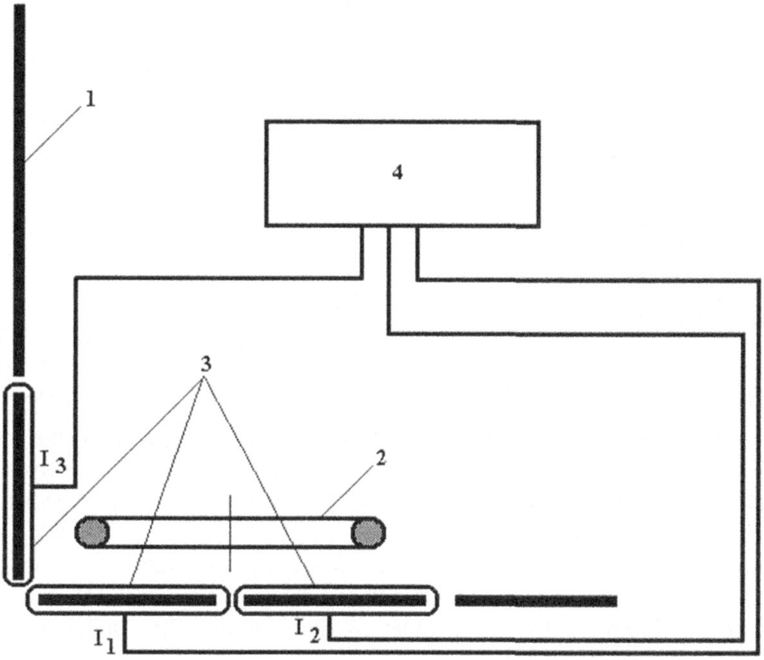

Fig. 2.19 Experiment scheme for the induced currents measure. 1 is the bent sheet billet with holes; 2 is the inductor; 3 are the induction probes, i.e., "Rogovsky belts"; 4 is the oscillograph ($I_{1,2,3}$—are the measured currents)

The holes for the induction probes were fulfilled in the sheet billet. The distance between them was the same and equal to half the width of the rectangular-shape inductor ($\sim$0.01 m).

The induction probes were absolutely identical. It should be noted that it was necessary in order to formulate the measured results in terms relative values (Tables 2.1 and 2.2). The norm of the measured currents had the current amplitude in an area on the horizontal plane of bend under the "nearest" current-wire—I_1.

Let us comment on the measured results.

For comparison with the data for the magnetic field intensity, one should take into account that the currents induced in the different parts of the sheet billet are the integrals of the according surface-current densities along the closed circuit around the distinguished area.

Thus, the first line of Table 2.1 confirms the reliability of the measurement shown in Fig. 2.2. The second line confirms the reliability of the measurement in Fig. 2.3. Finally, the third line is the integrated dependence shown in Fig. 2.4.

The data in Table 2.2 illustrate qualitatively the reliability of curves for the space distribution of the magnetic field intensity shown in Fig. 2.5.

Table 2.1 Relative values of the currents induced by the single-turn inductor

I			
α	I_1	I_2	I_3
0°	1.0	1.17	0.77
15°	1.0	0.91	0.82
45°	1.0	0.1	1.0

Table 2.2 Relative values of the currents induced by the two-turn inductor (the serial electrical connection)

I			
α	I_1	I_2	I_3
15°	1.0	0.8	1.0

The main conclusions by fulfilled measurements Include the following:

1. In the metal of the horizontal plane of bend, which is coplanar to the inductor plane, the current is being excited by two local maximums of density situated under the inductor current-wires, at which the second maximum exceeds the first maximum by 1.25 times.
2. The change of the angle slope of the inductor plane leads a decrease of the second maximum of the excited current: Thus, under $\alpha \approx 15°$ the second maximum decreases by 0.3 relative to the first one, but under $\alpha \approx 45°$ it decreases to practically zero.
3. The principle concentration of the magnetic-pressure forces in the centre of the angle zone becomes possible in the inductor system with the single-turn solenoid of rectangular shape, the plane of which is located along the bisector line of the flat angle being formed with one of the lengthy current-wires being situated along the bend line.
4. The experimental test was realized for the crucially new construction of the inductor system consisting of two turns of the rectangular geometry, the lengthy current-wires of which are joint along the bend line of the metal sheet, which allows to concentrate the field and pressure in the working zone as well to increase the effectiveness of the force action at the expense of the serial connection to the power source (the intensity amplitude increases by 2 times and the pressure amplitude by 4 times).

2.7 Inferences Gleaned from this Chapter

The main results of the investigations fulfilled by the theme of the present book, the first part are, generalized in the next statements.

1. Theoretical analysis of the electrical dynamics processes in the inductor systems' different options destined for "filling in" the angles in the bent metal sheet bent is performed.

2. The analytical expressions for the magnetic-field intensities in the suggested options of the inductor systems were obtained, and their reliability was confirmed by ultimate passing to both the known and proposed solutions.
3. As was shown by calculations, the current flowing through the parallel thin current-wires of the rectangular turn, which was placed in the inner cavity of the plane angle of the bent metal sheet with high specific electrical conductivity, is exciting on the coplanar planes the intensive magnetic fields and the great pressure forces with the same time shape. The intensity reaches $\sim 10^7$ A/m for the character quantity of current ~ 100 kA in the magnetic pulsed metal processing. The pressure maximums (up to ~ 100 MPa) of the space distribution will be localized in points with coordinates according to arrangements of the current-conductors.
4. Fulfilling the current wire with finite width essentially decreases the amplitudes of the excited fields and pressures. Thus, on passing from the point to the strip model in the calculations, the maximums of the magnetic field intensity decreases by approximately 33%, whereas the pressure was fixed at approximately 89%. The given fact was easily explained: The point model and, consequently, the greatest intensities occur when the correlation of the width and the distance up to the conducting plane moves toward zero.
5. There exist two maximums under the current-wires. However, in the case of current-wires with finite width, the excess of the second maximum above the first one, as for intensity and pressure, will be essentially lower than in the case of thin conductors.
6. The combination of the plane-slope angle where the current wires are situated and the width of the "far" current wire permits realizing the concentration of a high level of field and pressure in the angle zone of the bent metal sheet bent.
7. For the optimum option of the inductor system with two separate turns of the rectangular geometry that are inputted serially to the power source (the magnetic pulsed installation), the excited force is directed to the centre of the bend of the metal sheet. The amplitude of this force can reach tens of MPa.
8. In the metal of the horizontal plane of the bend, which is coplanar to the inductor plane, the current is excited with two local maximums of density, which are situated under current-wires of inductor, and the second maximum exceeds the first one by 1.25 times.
9. The change of the angle slope of the inductor plane leads to a decrease of the second maximum of the excited current: Thus, under $\alpha \approx 15°$, the second maximum decreases by 0.3 relative to the first one, but under $\alpha \approx 45°$ it decreases to practically zero.
10. The principle concentration of the magnetic pressure forces in the centre of the angle zone becomes possible in the inductor system with the single-turn solenoid of rectangular shape, the plane of which is located along the bisector line of the flat angle being formed with one of the lengthy current-wires being situated along the bend line.

11. Experimental testing was performed for the principle new construction of an inductor system consisting of two turns of rectangular geometry, the lengthy current-wires of which are made joint along the bend line of the metal sheet, which allows it to concentrate the field and pressure in the working zone as well increase the effectiveness of the force action at the expense of the serial connection to the power source (the intensity amplitude increases by 2 times and the pressure amplitude by 4 times!).

References

Altynova, M. M., & Daehn, G. S. (1977). English title: Electromagnetic metal forming handbook of I. Belyy, S. Fertik, L. Khimenko 1996 (The Russian Edition, 1977).

Austin, E. R., & Davies, R. (1970). *Developments in high speed metal forming* (American ed., pp. 184–271). New York: Industrial Press Inc.

Batygin, Y. V., Chaplygin, E. A., & Serikov, G. S. (2007). Inductor for the angles magnetic pulsed forming in the bent sheet-metal billets. Patent of Ukraine 31309

Batygin, Y. V., Golovashchenko, S. F., & Serikov, G. S. (2006). Magnetic fields and pressure being excited by the single turn inductor in the angle bent of the sheet billet [Russian]. Electrical Technique and Electrical Mechanics. Kharkov. Ukraine. № 6, pp. 66–71.

Batygin, Y. V., Khimenko, L. T., & Lavinski, V. I. (2003). The pulse magnetic fields for advanced technologies, vol. 1 [Russian]. The second edition re-made and supplemented. By general editing Doctor of Science, Professor Batygin, Y. V. Ukraine, Kharkov: Publishing House "Most-Tornado." 284 p.

Batygin, Y. V., Khimenko, L. T., & Lavinskiy, V. I. (2004). Direction change of the force action upon conductor under frequency variation of the acting magnetic field. In *Proceedings of the 1st International Conference on High Speed Forming*, Dortmund, pp. 157–160.

Bell, J. S., Kearsley, M. J., & Sykes, J. B. (1982). 2nd Edition revised and enlarged by E. M. Lifshitz and L. P. Pitaevskii. Oxford, New York, Beiling, Frankfurt: Pergamon Press, 620 p.

Daehn, G. S., Shang, J., & Vohnout, V. J. (2004). Improved formability by controlled of strain distribution in sheet stamping using electromagnetic impulses. In *Proceedings of the 1st International Conference on High Speed Forming*. Dortmund, Germany, pp. 211–220.

Knoepfel, H. (1970). *Pulsed high magnetic fields* (pp. 34–36). Amsterdam, London: North-Holland Publishing Company.

Korn, A., & Korn, M. (1968). *Mathematical handbook for scientists and engineers* (832 p.). Second, enlarged and revised edition. New York. San Fransisco, Toronto, London, Sidney: McGraw-Hill Company

Landau, L. D., & Lifshitz, E. M. (1984). *Electrodynamics of continuous media* [translated from Russian], pp. 105–113, 126–129, 199–210, v. 8 of Course of Theoretical Physics.

Psyk, V., Risch, D., Kinsey, B. L., Tekkayaa, A. E., & Kleiner M. (2011). Electromagnetic forming—A review. *Journal of Materials Processing Technology, 211*, 787–829.

Shneerson, G. A. (1981). Polja i perekhodnye processy v apparature sverkhsilnykh tokov [Russian]. USSR, Leningrad: Publishing House "Energia." 199 p.

Chapter 3
Magnetic Pulsed Attraction of Thin-Walled Metals

This section describes two fundamentally new ways of achieving the attraction of metals with both ferrous and non-ferromagnetic properties. Analytical solutions describing the inductor systems' operation of a given geometric shape show the expected operability of the system. The description of detailed mathematical calculations can be useful for further engineering calculations in the manufacture and modernization of such systems as EMF. Thus, when certain frequencies of the system's operating conditions are realized, the transition from the known effect of magnetic pressure to the effect of sheet-metal attraction is also realized. In turn, the improvement in the design of the inductor system makes it possible to attain the attraction of non-magnetic sheet metals. The optimal geometric shapes of inductor tools obtained during calculations are confirmed experimentally under non-factory laboratory conditions.

All well-known "traditional" EMF-applications are based on repelling Lorentz forces between the EMF coil and the conductive blank. In such a configuration, the coil and the tool (e.g., the forming die, the mandrel to which the blank is welded or crimped, the shearing edge used to cut the blank) are positioned at opposite sides of the blank. There is another configuration of EMF processes where the blank is attracted to the coil. Factually, the question is about the new scientific direction in the magnetic pulsed processing of metals (EMF) when conditions are created for transformation of the natural repelling action of conductors in their attraction by the pulsed magnetic field.

First, the main particularities of the suggested material not only enumerates the last achievements of the new scientific direction, it illuminates the main works with theoretic–experimental foundations of the magnetic pulsed attraction of thin-walled sheet metals. At that, the author's previous publications about the covered theme are being cited and even reprinted in full.

Finally there is a last methodical remark: This chapter consists of two subparts (3.1 and 3.2), each of which contains a separate numeration of the formulas and pictures that is accepted as beginning from Fig. 3.1.

Y. Batygin et al., *Electromagnetic Metal Forming for Advanced Processing Technologies*, Materials Forming, Machining and Tribology,
https://doi.org/10.1007/978-3-319-74570-1_3

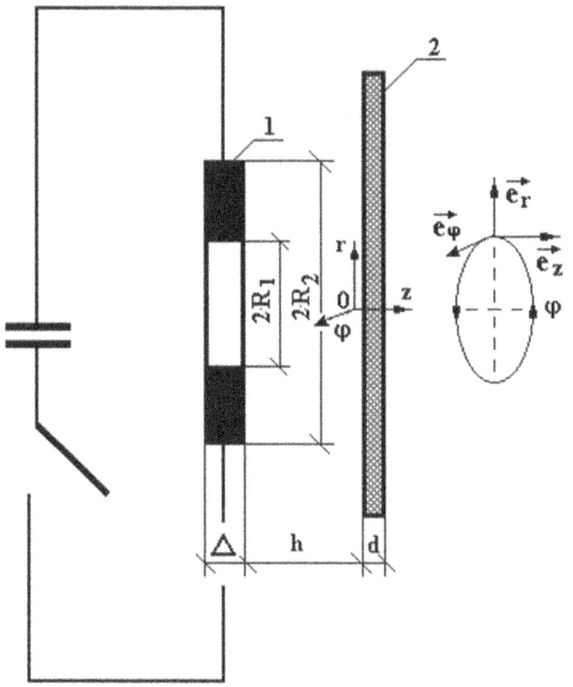

Fig. 3.1 Schematic of a single-turn inductor (1) interacting with a ferromagnetic sheet-metal blank (2) where $\vec{e}_r, \vec{e}_\varphi, \vec{e}_z$—are the unit vectors in the cylindrical coordinate system

3.1 The Attraction of Ferromagnetic Metals

Attracting forces are always present if magnetic fields interact with bodies consisting of ferromagnetic material. However, they are approximately one to two levels of magnitude smaller than those repelling Lorentz forces arising in typical electromagnetic forming processes at typical frequencies. However, in electromagnetic attraction processes, the magnetic forces play a very important role and must be taken into account.

The concept of a single-frequency sheet-metal attraction was introduced by Batygin et al. (2004, 2006) for ferromagnetic materials. This concept was discovered during EMF experiments with steel sheet-metal material deformed by electric discharges with different frequencies.

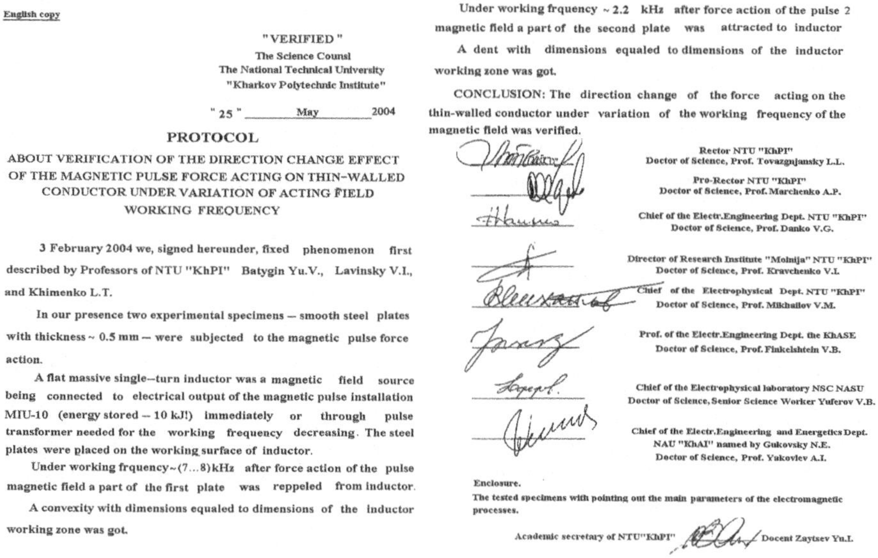

English copy

"VERIFIED"
The Science Counsl
The National Technical University
"Kharkov Polytechnic Institute"

"25" May 2004

PROTOCOL

ABOUT VERIFICATION OF THE DIRECTION CHANGE EFFECT OF THE MAGNETIC PULSE FORCE ACTING ON THIN-WALLED CONDUCTOR UNDER VARIATION OF ACTING FIELD WORKING FREQUENCY

3 February 2004 we, signed hereunder, fixed phenomenon first described by Professors of NTU "KhPI" Batygin Yu.V., Lavinsky V.I., and Khimenko L.T.

In our presence two experimental specimens – smooth steel plates with thickness ~ 0.5 mm – were subjected to the magnetic pulse force action.

A flat massive single–turn inductor was a magnetic field source being connected to electrical output of the magnetic pulse installation MIU-10 (energy stored – 10 kJ!) immediately or through pulse transformer needed for the working frequency decreasing. The steel plates were placed on the working surface of inductor.

Under working frquency~(7...8) kHz after force action of the pulse magnetic field a part of the first plate was reppeled from inductor.

A convexity with dimensions equaled to dimensions of the inductor working zone was got.

Under working frquency ~ 2.2 kHz after force action of the pulse 2 magnetic field a part of the second plate was attracted to inductor

A dent with dimensions equaled to dimensions of the inductor working zone was got.

CONCLUSION: The direction change of the force acting on the thin-walled conductor under variation of the working frequency of the magnetic field was verified.

Rector NTU "KhPI"
Doctor of Science, Prof. Tovazgnjansky L.L.

Pro-Rector NTU "KhPI"
Doctor of Science, Prof. Marchenko A.P.

Chief of the Electr.Engineering Dept. NTU "KhPI"
Doctor of Science, Prof. Danko V.G.

Director of Research Institute "Molnija" NTU "KhPI"
Doctor of Science, Prof. Kravchenko V.I.

Chief of the Electrophysical Dept. NTU "KhPI"
Doctor of Science, Prof. Mikhailov V.M.

Prof. of the Electr.Engineering Dept. the KhASE
Doctor of Science, Prof. Finkelshtein V.B.

Chief of the Electrophysical laboratory NSC NASU
Doctor of Science, Senior Science Worker Yuferov V.B.

Chief of the Electr.Engineering and Energetics Dept.
NAU "KhAI" named by Gukovsky N.E.
Doctor of Science, Prof. Yakovlev A.I.

Enclosure.
The tested specimens with pointing out the main parameters of the electromagnetic processes.

Academic secretary of NTU"KhPI" Docent Zaytsev Yu.I.

The suggested concept quickly found its practical application in dent-removal devices for automotive applications where the majority of exterior panels are fabricated from low-carbon steel (Web site of the Swiss company Beulentechnik AG).

The single-frequency experiments initially discussed in Batygin et al. (2004, 2006) for ferromagnetic materials and indicated that the direction of the electromagnetic pressure for ferromagnetic steel sheet-metal blanks is a function of the frequency of the discharge: At some limit, the applied electromagnetic force changed its direction: For high frequencies, repelling took place, but for low frequencies attraction was displayed. The effect of attraction is possible only for low frequencies, when

$$\omega \ll \frac{1}{\mu \cdot \gamma \cdot d^2} \tag{3.1}$$

where $\omega = 2\pi \cdot f, f$—is the working frequency; μ—is the permeability of metal; γ—is electric conductivity of sheet metal; and d—is sheet-metal thickness.

It should be understood that formula (3.1) provides only the upper limit of the frequency at which attraction can be expected. The actual boundary frequency between attraction and repelling can be lower. Physically, formula (3.1) can be further explained in the following way: It compares the sheet-metal thickness with the magnetic-diffusion length at the given frequency. It gives a rough estimate, if the magnetic field's energy remains in the work piece, where it gives rise to eddy currents and hence invokes Lorentz forces, or is mainly transferred through them. However, without an estimate on the size of magneto-static forces, it cannot be used as an indicator whether Lorentz-forces or magneto-static forces prevail. Obviously, formula (3.1) provides a necessary but insufficient condition of electromagnetic attraction. An analytical method defining the boundary frequency between repelling and attraction will be discussed further.

3.1.1 Theoretical Analysis of the EMF-Attraction Process for Ferromagnetic Metals

The interaction between a single-turn inductor and a ferromagnetic sheet-metal blank was analyzed. The schematic of the process is shown in Fig. 3.1. The following assumptions were made in order to simplify the mathematical model of the process:

1. The single-turn inductor is sufficiently thin (its thickness is $\Delta \to 0$) and "transparent" for the acting fields, so its metal has no effect on the considered electromagnetic processes.
2. The process is considered quasi-stationary according to the criterion $\frac{\omega}{c} \cdot \ell \ll 1$ where ω—is the cyclic frequency of the process; c—is the light velocity for vacuum; and l is the greatest typical dimension of the system under consideration. Physically it means that the typical dimension in the system is substantially smaller than the wavelength of the acting electromagnetic field.

1. The system has an axial symmetry, so $\partial/\partial\varphi = 0$ (φ—is the polar angle).
2. The sheet-metal blank is sufficiently thin, and its radial dimension is sufficiently large, so $d/R_{1,2} \ll 1$ and $\omega \cdot \tau \ll 1$ ($\tau = \mu_1 \cdot \gamma \cdot d^2$, γ, d—are conductivity and thickness of the sheet-metal blank, respectively, and $R_{1,2}$—are the internal and external radii of the inductor, respectively).
3. The permeability of the sheet-billet metal is constant and equal to $\mu_1, \mu_1 = \mu_0\mu_r, \mu_0$ and μ_r—are, respectively, the vacuum permeability and the relative permeability of the metal.
4. Reproduction of the mutual inductance of tool coil and work-piece is neglected.
5. The sheet metal's motion is not taken into account.
6. The nonlinear behavior of ferromagnetic saturation is neglected considering an average value of magnetic permeability.
7. The variability of conductivity as a consequence of temperature increase due to Joule heating as well as magnetization is neglected.
8. The reversion of magnetization is not taken into account.

The Maxwell equations in the L—space for the electromagnetic-field components being excited with zero initial conditions ($E_\varphi \neq 0, H_{r,z} \neq 0$) can be presented in the following well-known form:

$$\begin{cases} \dfrac{\partial H_r(p,r,z)}{\partial z} - \dfrac{\partial H_z(p,r,z)}{\partial r} = j_\varphi(p,r,z); \\ 1/r \cdot \partial/\partial r\left(r \cdot E_\varphi(p,r,z)\right) = -\mu \cdot p \cdot H_z(p,r,z); \\ \dfrac{\partial E_\varphi(p,r,z)}{\partial z} = \mu \cdot p \cdot H_r(p,r,z); \end{cases} \tag{3.2, 3.3, 3.4}$$

where p—is the parameter of the Laplace transformation;

$$E_\varphi(p,r,z) = L\{E_\varphi(t,r,z)\}; H_{r,z}(p,r,z) = L\{H_{r,z}(t,r,z)\};$$
$$j_\varphi(p,r,z) = L\{j_\varphi(t,r,z)\}$$

The current density in the right part of Eq. (3.2) can be represented as

$$j_\varphi(p,r,z) = (p \cdot \varepsilon_0 + \gamma) \cdot E_\varphi(p,r,z) + j_{\varphi\, i}(p,r,z), \tag{3.5}$$

where

$j_{\varphi\, i}(t,r,z)$ is the current density in the inductor,

$$j_{\varphi\, i}(p,r,z) = j(p) \cdot f(r)\delta(z+h), \quad j(p) = \frac{I(p)}{R_2 - R_1}, I(p) = L\{I(t)\};$$

$f(r)$ is the radial distribution function for the current in the inductor turn;
$\delta(z+h)$ is the Delta-function;and
μ_a is the permeability of the environment, $\mu_a = \begin{cases} \mu_0, \text{the vacuum,} \\ \mu_1, \text{the ferromagnetic} \end{cases}$.

To solve the formulated problem, the areas with homogeneous electrical physical characteristics must be separated. Such areas include air-non-conductor ($z \in (-\infty, 0]$) sheet metal blank–metal ($z \in [0, d]$), and air-non-conductor ($z \in [d, +\infty)$). In each separated area, the differential equation system (3.2) through (3.4) must be integrated taking into account the equation for the current density (3.5). The axial symmetry of the system being considered—as well as the requirement of the limit value for the electromagnetic field intensity for $r = 0$ and $r = \infty$—allow the use of the integral Fourier Bessel transformation described by Korn and Korn (1968).

The general integrals of differential Eqs. (3.2) through (3.4) include integration constants, which can be defined from continuity condition for tangential components of the electromagnetic-field intensity at the boundaries of the selected regions. Finally, the obtained equation for the system under consideration is defined by the mutual influence of fields in the selected regions. The solution of the problem formulated previously can be written in the form of expressions for the currents and the fields being excited in a ferromagnetic sheet-metal blank as shown in Fig. 3.1, $z \in [0, d]$.

The current density:

$$j_{\varphi}(z,r,\psi) = 2 \cdot j_m/d \cdot \int_0^{\infty} f(x) \cdot \exp(-x \cdot h/d) \cdot x \cdot J_1(x \cdot r/d)$$
$$\cdot \sum_{k=0}^{\infty} \frac{\beta_k \cdot F_{1k}(\mu_r \cdot x, z)}{\Phi_k(\mu_r \cdot x)} \times \frac{\mathrm{d}j(\psi)}{\mathrm{d}\psi} * \exp\left(-\frac{(\beta_k^2 + x^2)}{\omega\tau} \cdot \psi\right) \mathrm{d}x \qquad (3.6)$$

where

j_m and $j(\psi)$ — are the amplitude of the current density in the inductor its time function,

$\psi = \omega \cdot t$ — is the phase,

ω — is the cyclic frequency of the exciting current,

$f(x) = 1/d^2 \cdot \int_{R_1}^{R_2} f(r) \cdot r \cdot J_1(x \cdot r/d) \cdot dr, f(r)$ — is the radial distribution of the current in the inductor,

$J_1(x \cdot r/d)$ — is the first-order Bessel function,

β_k — are the roots of the equation: $\mathrm{tg}(\beta_k) = -\frac{2}{\frac{\beta_k}{\mu_r \cdot x} - \frac{\mu_r \cdot x}{\beta_k}}$,

$$F_{1k}((\mu_r \cdot x), z) = (\mu_r x) \cdot \sin(\beta_k(1 - z/d)) + \beta_k \cdot \cos(\beta_k(1 - z/d)),$$
$$\Phi_k(\mu_r \cdot x) = \cos(\beta_k)\left[(\mu_r \cdot x)^2 + 2(\mu_r \cdot x) - \beta_k^2\right] - 2\beta_k \sin(\beta_k)[1 + (\mu_r \cdot x)],$$

x—is the variable of integration.

The existence of the obvious limit transitions in the convolution of two functions in expression (3.6) should be emphasized.

$$\frac{\mathrm{d}j(\psi)}{\mathrm{d}\psi} * \exp\left(-\frac{(\beta_k^2 + x^2)}{\omega \cdot \tau} \cdot \psi\right) = \int_0^{\varphi} \frac{\mathrm{d}j(\zeta)}{\mathrm{d}\zeta} \cdot \exp\left(-\frac{(\beta_k^2 + x^2)}{\omega \cdot \tau} \cdot (\psi - \zeta)\right) \mathrm{d}\zeta$$
$$= \begin{cases} j(\psi), \gamma \to \infty, \\ \frac{\mathrm{d}j(\psi)}{\mathrm{d}\psi}, \gamma \to 0. \end{cases} \qquad (3.7)$$

According to the known conditions for the skin-effect regime ($\gamma \to \infty$), the time functions of the induced current and the inductor current are the same. For intensive penetration of the field ($\gamma \to 0$) regime, the induced current is proportional to the exciting current derivative.

The components of the magnetic-field intensity excited inside the blank can be defined by the following equations:

(a) the radial component (this is the tangential one),

$$H_r(z,r,\psi) = -\frac{2\cdot j_m}{(\omega\cdot\tau)}\int\limits_0^\infty f(x)\cdot\exp\left(-x\cdot\frac{h}{d}\right)\cdot x\cdot J_1(x\cdot r/d)\sum_{k=0}^{\infty}\left(\frac{\beta_k^2\cdot F_{2k}(\mu_r\cdot x),z)}{\Phi_k(\mu_r\cdot x))}\right.$$
$$\left.\times j(\psi) * \exp\left(-\frac{(\beta_k^2+x^2)}{\omega\cdot\tau}\cdot\psi\right)\right)\mathrm{d}x \tag{3.8}$$

where $F_{2k}((\mu_r\cdot x),z) = (\mu_r\cdot x)\times\cos(\beta_k(1-z/d)) - \beta_k\cdot\sin(\beta_k(1-z/d))$,

(b) the z-component (this is the normal one),

$$H_z(z,r,\psi) = -\frac{2\cdot j_m}{(\omega\cdot\tau)}\int\limits_0^\infty f(x)\cdot\exp(-x\cdot h/d)\cdot x^2\cdot J_0(x\cdot r/d)\sum_{k=0}^{\infty}\left(\frac{\beta_k\cdot F_{1k}(\mu_r\cdot x),z)}{\Phi_k(\mu_r\cdot x))}\right.$$
$$\left.\times j(\psi) * \exp\left(-\frac{(\beta_k^2+x^2)}{\omega\cdot\tau}\cdot\psi\right)\right)\mathrm{d}x \tag{3.9}$$

where $J_0(x\cdot r/d)$—is the zero-order Bessel function.

The obtained formulas (3.6), (3.8), (3.9) describe the distributions of the electric current and the magnetic field in the work-piece as a function of the exciting current. The space-time function of the magnetic pressure acting on the conducting ferromagnetic sheet metal can be written in the following well-known form:

$$P_L(r,t) = -\mu_1\cdot\int\limits_0^d j_\varphi(z,r,t)\cdot H_r(z,r,t)\cdot\mathrm{d}z. \tag{3.10}$$

Formula (3.10) added with expressions (3.6) and (3.8) is the solution of the problem about the Lorentz forces acting in the vector-normal direction to the surface of the blank.

The surface density of the forces caused by the magnetic properties of a thin sheet-metal blank can be defined by the following well-known method, which is based on the energy balance for a sufficiently small volume of ferromagnetic metal in the magnetic field:

$$P_M = -(\mu_0\cdot\mu_r)/2\cdot(\mu_r-1)\cdot\left[\left(H_{1z}^2-H_{2z}^2\right)+1/\mu_r\cdot\left(H_{1r}^2-H_{2r}^2\right)\right], \tag{3.11}$$

where $H_{1,2z}$ and $H_{1,2r}$—are "z" and "r" components of the magnetic-field intensities on the boundary surfaces of the blank facing the inductor (index—1) and on the opposite side of the blank (index—2).

The surface density of forces has a similar sense as pressure: In the case of electromagnetic attraction, it can be considered as a negative pressure.

Analysis of expression (3.11) shows the following:

- The magnetic metal sheet is subjected to the force proportional to the difference between the squares of the normal and the tangential components of intensities on the boundaries of the sheet metal. This force acts in the direction of the larger amplitude of the magnetic field unlike the Lorentz forces, which have the opposite direction.
- In the regime when $\mu_r \gg 1$, the acting force is determined by the normal components of the magnetic field.

The formulated hypothesis of the nature of change of the direction of forces acting on the ferromagnetic sheet metal can be quantified by adding the Lorentz force (3.10) and the force (3.11) caused by the magnetic properties of the sheet-metal blank being deformed.

$$P_s(r,t) = \mu_1\left(\int\limits_0^d j_\varphi(z,r,t)\cdot H_r(z,r,t)\mathrm{d}z - \frac{(\mu_r-1)}{2}\cdot\left[\left(H_{1z}^2 - H_{2z}^2\right) + \frac{1}{\mu_r}\cdot\left(H_{1r}^2 - H_{2r}^2\right)\right]\right) \tag{3.12}$$

Expression (3.12), along with formulas (3.6), (3.8), and (3.9), is the full solution of the stated problem.

3.1.2 Numerical Estimates

The numerical calculations were performed for an experimental inductor system with the following parameters: single-turn coil—$R_1 = 0.025$ m, $R_2 = 0.03$ m; and working gap between the coil and the sheet-metal blank—$h = 0.0005$ m. The current distribution in the coil along the radial coordinate was approximated by the following function: $f(r) = \eta(r - R_1) - \eta(r - R_2)$.

In the conducted experiments, the time function of the inductor-current density has the shape of an exponentially relaxing sine, that is, $j(t) = j_m \cdot \mathrm{e}^{-\delta_0\cdot\psi}\cdot\sin\psi$, where $j_m = I_m/(R_2 - R_1)$—is the amplitude of the current density; I_m—is the current amplitude; δ_0—is the relative decrement of relaxing; $\psi = \omega\cdot t$—is the phase; and ω—is the cyclic frequency.

The electrical conductivity of sheet-metal blank was assumed to be $\gamma = 0.4\times10^7$ 1/(Ohm m), and the sheet-metal thickness was $d = 0.00075$ m. The relative permeability value was taken as $\mu_r \approx 2.5$. Such a choice was based in details by Batygin et al. (2013).

The electric current amplitude in the inductor was $I_m = 50\,\text{kA}$, and the frequency in the pulse was $f = 2 \div 8\,\text{kHz}$. The relative damping coefficient was calculated as the damping coefficient divided by the cyclic frequency and was equal to $\delta_0 = 0.3$.

The excited forces were calculated using formula (3.10) for the distributed Lorentz forces P_L (known as "magnetic pressure"); the distributed forces of the magnetic attraction P_M (the analogue of the "magnetic pressure" but in the opposite direction) were defined based on formula (3.11); the distributed total forces acting on the ferromagnetic metal sheet in the magnetic field P_S were calculated based on formula (3.12). Results of calculations are presented in Figs. 3.2 through 3.4. Phase functions for distributed forces $P_L(\psi)$ and $P_M(\psi)$ are provided for radial coordinates where they have maximum values: $P_L(\psi)$ for $r \approx 0.9{\cdot}R_2$ and $P_M(\psi)$ for $r \approx 0.85{\cdot}R_2$ (Fig. 3.3).

A special comment should be made regarding the negative values of distributed Lorentz forces. For typical electromagnetic-forming processes, distributed Lorentz forces are usually repelling, which might raise some questions how they can become attractive at certain moments of the process being studied in this paper. The answer lies in the substantial difference of frequency of discharge in traditional electromagnetic-formation processes and the process studied in this paper. In traditional electromagnetic formation, a substantially higher frequency of discharge is employed to avoid electromagnetic-field diffusion through the thickness of the blank. Belyy et al. (1977) recommends selecting the coil and pulse generator parameters to satisfy this requirement of field non-penetration. In this case, the ideal

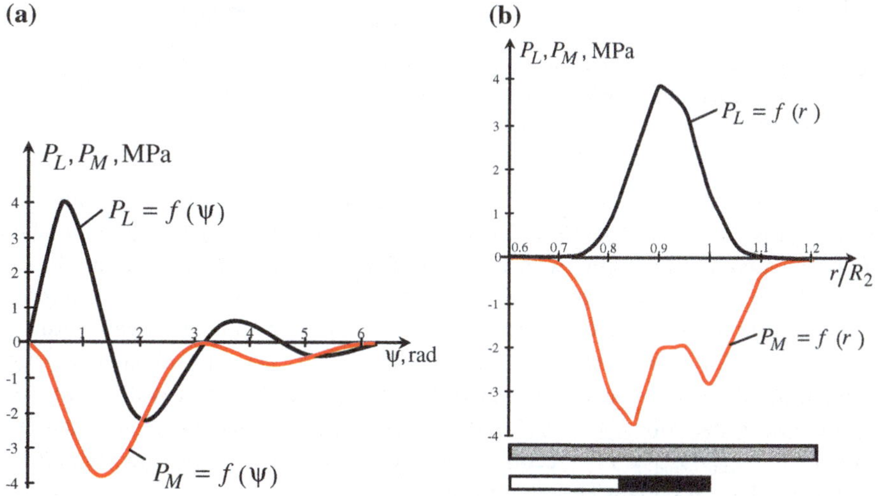

Fig. 3.2 Distribution of the distributed Lorentz forces P_L and distributed forces of the magnetic attraction P_M excited in the ferromagnetic sheet-metal blank by an approximately 2-kHz discharge. **a** The phase function. **b** The distribution along the radial coordinate

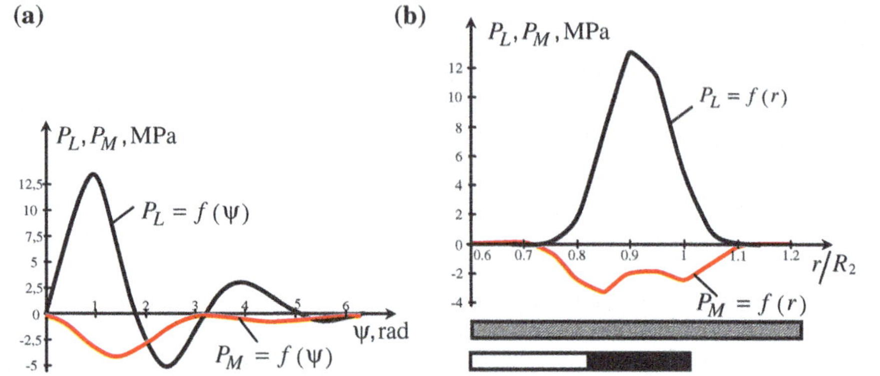

Fig. 3.3 Distribution of forces excited in ferromagnetic sheet-metal blank by approximately 8 kHz. **a** The phase function. **b** The distribution along the radial coordinate

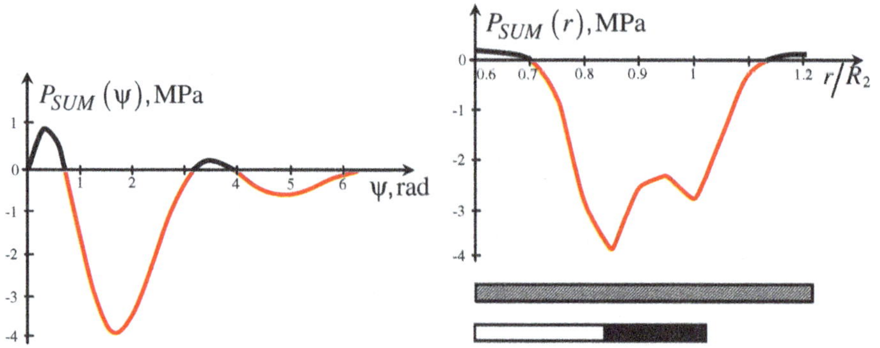

Fig. 3.4 Resulting distributed force excited in a ferromagnetic sheet-metal blank by approximately 2 kHz. **a** The phase function. **b** The distribution along the radial coordinate (the repelling forces are greater then zero, and the attracting forces are less than zero)

model of interaction of an external electromagnetic field of the coil with ideal conductor is valid. The current in the coil (I_{coil}) and the current induced in the blank (I_{ind}) are identical functions with a time shift of $\sim\pi$ and have equal amplitudes $I_{\text{coil}} \approx I_{\text{ind}} = I$. Therefore, the Lorentz-distributed force taken into consideration as an interaction of two opposite electrical currents (Ampere's law) is always positive: $P \sim I_{\text{coil}} \cdot I_{\text{ind}} \approx I^2 > 0$. However, if penetration of the electromagnetic field through the thickness occurs, the function of the induced current is different from the current in the coil, $I_{\text{coil}} \neq I_{\text{ind}}$, and the Lorentz-distributed force—defined as a product $P \sim I_{\text{coil}} \cdot I_{\text{ind}} \neq I^2$—becomes the oscillating function of time, which has negative values at certain periods of time.

It should be emphasized that even though the electromagnetic-attraction process must be substantially slower than the traditional electromagnetic-forming process, it is still a dynamic process with a load duration of <500 μs. Although the electromagnetic integrated attraction force increases with a decrease of the frequency and decrease of the Lorentz force, it should be stressed that further decrease of the frequency to a direct current process is not feasible. According to our experimental data and analytical calculations, the electric current required to produce dent removal is in the range of 40 kA, whereas the dynamic force is approximately 50% greater than the estimated quasi-static pressure (4 vs. 2.4 MPa). Obtaining the direct current value at the level of 40 kA is not technically feasible because such a process will consume the power of the whole district. Therefore, the charging-and-discharging sequence is really necessary to accumulate the electric energy inside the capacitors and then quickly discharge it to the high-voltage circuit.

A comparison of the calculation results permits us to make some preliminary conclusions.

- For low frequency, approximately 2 kHz, the integral action (in time) of magnetic forces on the ferromagnetic thin sheet-metal blank prevails over Lorentz forces, and, as a result, the sheet-metal blank is attracted to the inductor.
- When increasing the frequency up to approximately 8 kHz, the Lorentz forces increase by more than factor of 3, whereas the magnetic forces remain unchanged. The integral action of the Lorentz forces prevail, whereas the attraction forces decrease: The electromagnetic repelling pressure on the sheet-metal blank takes place.

The obtained results provide an explanation of the physical nature of attraction forces, which were demonstrated by decreasing the frequency of the discharge. Increasing the frequency of the discharge caused repelling pressure to dominate over attracting forces. Conducted analysis of repelling and attracting forces indicated that the following:

- The maximums of attracting and repelling forces shift in time as shown in Fig. 3.2a: The amplitude of the magnetic attracting force occurs at $\varphi_{\max,M} \approx 1.256\,\text{rad}$ independent of frequency. The repelling forces achieve their maximum at $\varphi_{\max,L} \approx 0.628\,\text{rad}$ for 2 kHz and at $\varphi_{\max,L} \approx 0.942\,\text{rad}$ for 8 kHz, which is caused by the different "shape" of the function f of the induced currents versus time.
- the maximums of attracting and repelling forces are applied to different areas as shown in Fig. 3.2b: The radial maximum of the repelling forces are located under the center of the turn, and the attracting forces achieve their maximum near the edge of the turn.

Based on these considerations, it is beneficial to add the calculations of the integral force as a function of the working frequencies. It should be noted that in previous equations, the distributed force has the character of pressure, whereas in the equation below, the pressure is integrated along the surface of the blank and therefore has a character of force. The resulting force was calculated through the integration of pressure through the radial coordinate and time and then averaging its value through the time period of the acting field. The results of calculations are presented in Fig. 3.5.

According to Fig. 3.5, decreasing the discharge frequency below the critical value of ~6 kHz leads to a change of direction of the integrated force: Below the marked critical frequency, thin-walled sheet ferromagnetic metal is attracted to the inductor, and above the frequency of ~6 kHz, the sheet-metal blank is repelled.

This approach is really an engineering estimate convenient for practical calculations. It attempts to replace the integration of a solid-mechanics equation of motion coupled with a numerical solution of Maxwell equations. Such a numerical approach provides the oscillations of the sheet-metal blank driven by attracting magnetic forces and repelling/attracting Lorentz forces. Obviously, the simplified approach can never fully replace the numerical solution of the problem with substantially less number of assumptions. Comparing the results of the numerical solution and a simplified analytical solution would be an interesting development of this work for the future.

In addition, it should be emphasized that for EMF processes employing ferromagnetic sheet-metal blanks of low electrical conductivity with relatively slow electric discharges, the magnetic forces should be taken into account in order to achieve a correct representation of the electromagnetic forces applied to the blank.

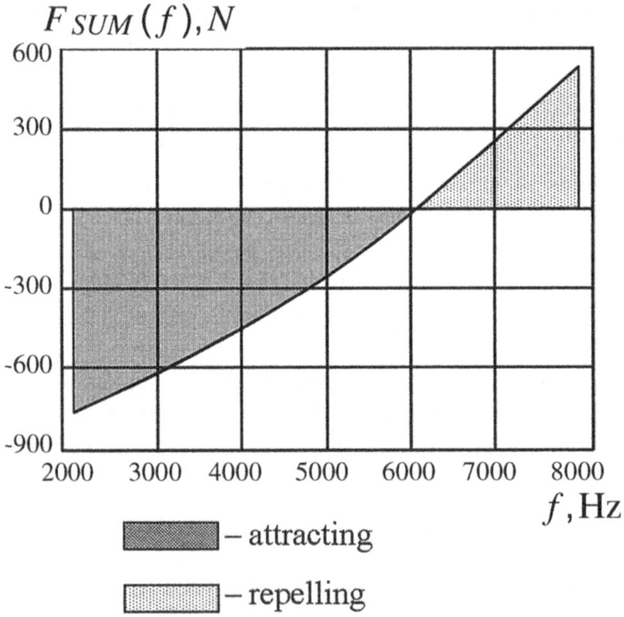

Fig. 3.5 Average integral force acting on a thin-walled ferromagnetic sheet as a function of the frequency of the acting field

Calibration of advanced high-strength steels with EMF processes, as discussed by Iriondo et al. (2011), provides one of such example.

3.1.3 Experimental Validation

Experiments were performed using a laboratory pulse generator (Fig. 3.6) capable of producing multiple pulses repeating with a frequency of 3.5 Hz. The capacitors were charged to 1800 V corresponding to accumulated energy in a single discharge of 2 kJ.

The pulse generator has the internal frequency $f_0 \approx 7\,\text{kHz}$, and the internal inductance is $L_0 \approx 430 \times 10^{-9}$ Hn. A single-turn massive inductor with an internal diameter of 40 mm was connected to the pulse generator by way of a current-pulse transformer with a coefficient of transformation of 5. The discharge frequency was measured as 1.9 kHz. The amplitude of electric current flowing through the coil was 38 kA. The deep-drawing quality (DDQ) of steel samples was 0.8-mm thick. The DDQ steel samples of the flat sheet were positioned next to the single-turn coil as shown in Fig. 3.7, which was insulated from the coil with an approximately 1-mm layer of insulation. Initially the flat samples were bulged by the attraction inside the window of the single-turn coil. The bulge was axisymmetric and had a maximum height of 1.5 mm at the symmetry axis of the coil.

Fig. 3.6 Laboratory pulse generator for EMF-attraction processes with maximum charging voltage of 2 kV and accumulated energy of 2 kJ capable of producing multiple pulses repeating with a frequency of 3.5 Hz. 1 is the pulsed current transformer; 2 are the sheet-metal samples; 3 is the pulse generator; and 4 is the control panel

Fig. 3.7 Single-turn coil employed in the EMF-attraction experiments for ferromagnetic sheet metals

Assuming that the bulge can be approximated as a spherical arc, the equivalent quasi-static pressure that would form such a bulge can be calculated using the well-known analytical formulas for a circular diaphragm. The radius of the spherical arc was calculated as $R = a \cdot \left(1 + (h/a)^2\right) \cdot a/2h$ where a is half of the diameter of the bulged circular diaphragm $a = 20$ mm; and $h = 1.5$ mm is the maximum height of the bulge. Based on these calculations, $R = 134$ mm. The quasi-static uniform bulge pressure was estimated for a flexible circular diaphragm as $p = 2 \cdot YS \cdot t/R$ where YS is the yield stress of 200 MPa; $t = 0.8$ mm; and R = 134 mm. The quasi-static attracting pressure can be estimated as $P = 2.4$ MPa. From this estimate, it is obvious that the electromagnetic-attraction process can develop substantially a larger attraction pressure than vacuum dent removal where pressure is limited to 0.1 MPa (atmospheric pressure). The described electromagnetic-attraction process provides an attraction-pressure factor 24 times greater than atmospheric pressure.

The proposed estimate of the quasi-static pressure necessary to form the bulge can be used as an engineering estimate of the necessary electromagnetic pressure. In Fig. 3.4, the amplitude of the attracting pressure was estimated as 4 MPa, which is an indication that some more bulging could be accomplished with such a level of attracting pressure.

After the initial bulge was produced on a flat blank using the described process of electromagnetic attraction, it was flipped over, and the dent-removal process was physically simulated by pulling the bulge back using an electromagnetic-attraction process. The dent-removal process resulted in a decrease of bulge height by a factor thereof 3. Further improvement of the process can be accomplished by decreasing the frequency of the discharge even lower and also by increasing the energy of the discharge. The sample of DDQ steel after producing the initial dent on the flat sheet by electromagnetic-attraction process is shown as N1 in Fig. 3.8. The result of dent removal, which was achieved by flipping the sample over and then removing the dent by repeating the attraction process, is termed "N2" in Fig. 3.8.

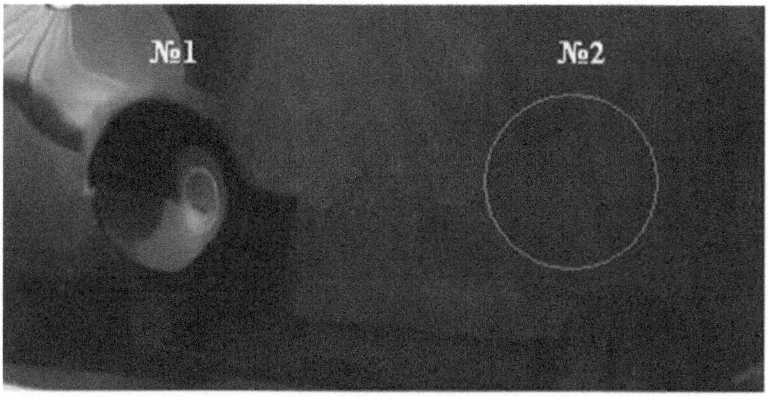

Fig. 3.8 Sample of DDQ-steel sheet after electromagnetic-attraction experiments. (1) Sample after producing a dent on the flat sheet using electromagnetic-attraction process. (2) Sample after removing the dent by repeating the attraction process after flipping the sample over

Conclusions

1. The suggested EMF-attraction process can deform ferromagnetic sheet-metal materials, such as low-carbon steels, using low-frequency discharges.
2. The analytical model, based on the solution of Maxwell equations, explained that the magnetic forces prevail over the Lorentz forces for low-frequency discharges.
3. For relatively slow EMF processes applied to ferromagnetic sheet-metal blanks of low electrical conductivity, the magnetic forces should be taken into account in order to achieve correct representation of the electromagnetic forces applied to the blank.

3.2 Pulsed Electromagnetic Attraction of Non-magnetic Sheet Metals

The objective of this section is to introduce a new manufacturing process of attracting non-magnetic sheet material through the recently developed EMF-tools including the single-turn coil and the so-called attracting screen. This process is one of the few examples of generating attracting forces with much higher capabilities than mentioned before vacuum forming. The first application of this process is expected to be in automotive repair shops where an attracting one-sided distributed force may enable the low-cost correction of exterior-panel damage.

3.2.1 Concept of Pulsed Electromagnetic Attraction of Non-magnetic Metals

Electromagnetic attraction of the designated areas of panels manufactured from sheet-metal blanks is motivated by the restoration needed by exterior panels of aircrafts and ground vehicles. Analysis of real-life examples indicates that a great number of types of damage are dents in areas where access from the inside of the panel is problematic or impossible. These cases include aircraft wings and fuselage components as well as doors, hoods, roofs, fenders, and bumpers of cars and trucks. From this perspective, there is a practical interest in developing devices allowing the restoration of sheet-metal parts by eliminating dents on the surface from the external side without disassembling the components, thus avoiding damage to the existing paint and protective coating. These systems include interruption of slowly increasing fields and exciting currents as well as the natural effects of changing the force direction applied to a thin ferromagnetic sheet using low-frequency electromagnetic processes. All mentioned concepts have certain disadvantages and limitations that complicate their practical usage such as employment of expensive switching components enabling the interruption of high-intensity currents and/or the requirement of the component material to be ferromagnetic.

For the first time, the initial concept of electromagnetic attraction based on creating unidirectional currents in the blank and in the attracting screen was formulated by Batygin et al. (2006). The physical concept of the proposed coil design is based upon the well-known Ampere's law. To create unidirectional parallel currents, an additional conductive screen was added to a traditional single-turn inductor design. A well-known single-turn coil design is employed to create attraction forces using a new process: The inductor is positioned between parallel plates of the screen and the sheet-metal blank requiring treatment by attracting forces. Identically directed eddy currents are induced in the metal of the screen and the sheet-metal blank by the current flowing through the single-turn coil. The proposed design requires a low-frequency discharge to accomplish the intensive penetration of the electromagnetic field through the thickness of the blank and avoid repelling Lorentz forces, which are the natural result of interaction of a magnetic field with a conducting environment.

To make the new concept clearer, it must be discussed in comparison with the concept of traditional EMF processes. As has been indicated in a large number of publications on the analyses and applications of traditional EMF processes, traditional EMF technology, as shown schematically in Fig. 3.9, requires a high frequency of discharge of capacitors through the coil in order to induce an electric current in the blank and contain the electromagnetic field within the skin layers of both the blank and the coil as well as the clearance between them.

Known methodologies of electromagnetic coil design and accepted design recommendations for EMF equipment typically intend to avoid penetration of the electromagnetic field through the thickness of the blank. The reason is that the pressure applied to the blank can be calculated as:

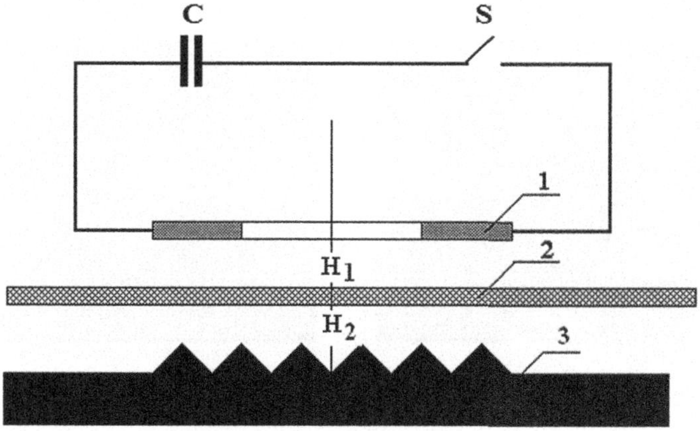

Fig. 3.9 Schematic of the traditional EMF process. 1 = single-turn inductor; 2 = sheet metal blank; 3 = one-sided die; C = block of capacitors for electric energy storage; S = switch of the discharge circuit; H_1 = electromagnetic field intensity in the clearance between the coil and the blank; and H_2 = intensity of the electromagnetic field penetrated through the blank

$$P = \frac{\mu_0}{2} \cdot \left(H_1^2 - H_2^2\right),$$

where H_1 is the electromagnetic field intensity in the clearance between the coil and the blank and H_2 is the intensity of the electromagnetic field penetrated through the blank.

Based on this consideration, the number of turns of the coil and the capacitance and voltage of the equipment is usually selected in addition to the general intention to keep the connection between the equipment and the coil at a minimum inductance level. In such a configuration, the current induced in the blank has almost the same time form (but with phase difference $\sim\pi$) as the discharge current propagating through the coil. The slight shift in time phases of currents in the coil and in the blank, caused by penetration processes and observable in the experiments, is usually not taken into account in making practical calculations of the EMF processes. In a traditional EMF configuration, a strong repelling electromagnetic pressure is observed between the coil and the blank with good electrical conductivity and high frequency of discharge.

Examples of experimental curves of exciting electric current running through the coil synchronized with the induced current running through the blank for the discharge frequency of 33 and 1.33 kHz are shown in Fig. 3.10a, b, respectively. Comparison of these two cases indicates that 33-kHz currents in the coil and in the blank have a phase shift of $\sim\pi$, which means that currents run in opposite directions resulting in a strong repelling force. For 1.33 kHz, the currents in the coil and in the blank have $\sim\pi/2$ shift in phases leading to very weak interaction between the currents, which periodically changes directions. This effect was fully discussed in

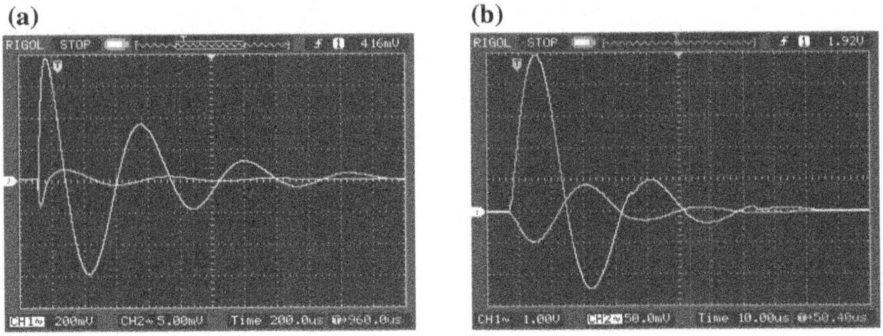

Fig. 3.10 Waveforms of pulsed electric current in the coil (higher amplitude curve) and in the blank (lower amplitude curve) for (left panel) a 33-kHz discharge current and (right panel) a 1.33-kHz discharge current

the analytical model by Batygin et al. (2013) for two ultimate cases when the electrical conductivity $\gamma \to 0$ and when $\gamma \to \infty$.

Technically, the EMF processes compete with traditional metal-forming operations in two-sided dies, quasi-static hydro forming processes, and super plastic forming processes based on hot-gas forming, other pulsed forming processes such as explosive forming, electro-hydraulic forming, and laser forming. All of these processes apply force directed outward from the tool to the blank.

The proposed process of electromagnetic attraction is among very few processes that can create force directed inward from the blank to the tool. A similar effect can be achieved by vacuum forming where the pressure differential is limited to the atmospheric pressure naturally applied outside of the blank and deforming it due to the absence of pressure inside the blank. It can be also compared with quasi-static electromagnetic force, which is appropriate only for magnetic materials.

According to the observed result, the concept of the process has some remote similarity to vacuum forming; however, the process is applied to electromagnetic forming. The idea is to create an electromagnetic field from both sides of the blank and then eliminate it by an opposite electromagnetic field generated by the screen. This process involves and additional piece of tooling: the electromagnetic screen, which plays a critical role in creating the electromagnetic field of the opposite sign and force interaction of conductors with unidirectional currents for attraction of the non-magnetic sheet metals.

To eliminate the repelling force between the coil and the blank, the discharge frequency must be rather slow to accomplish significant penetration of the electromagnetic field through the blank and also to increase the time shift in phases of current in the coil and in the blank: The induced current in the blank is now proportional to the derivative of the inducing current by time; for harmonic processes this shift in phases is $\sim \pi/2$. This second circumstance is ultimately important because it makes the dynamic interaction between the coil and the blank almost non-essential, whereas the screen starts playing a major role because the

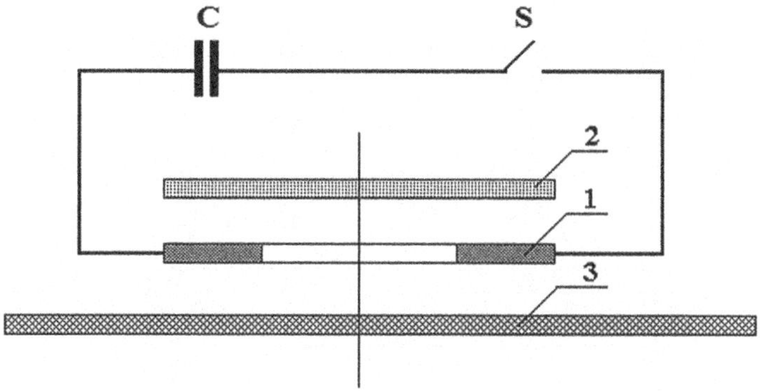

Fig. 3.11 Schematic of ISAS tool. 1 = single-turn inductor; 2 = accessory attracting screen; 3 = sheet-metal blank; *C* = block of capacitors for electric-energy storage; and *S* = switch of the discharge circuit

phases of electric current induced in the blank and in the screen are identical. In this case, the coil is not subjected to any significant electrodynamic forces contrary to traditional EMF processes.

The schematic of the proposed process of sheet-metal attraction, which was termed "inductor system with attracting screen" (ISAS) is shown in Fig. 3.11.

The tool ISAS operates in the following way: the preliminary charged-capacity storage (*C*) discharges through the circuit with the single-turn inductor (1) when the switch (*S*) is moved into the closed position by a special ignition circuit not shown in Fig. 3.11. The discharge current induces the unidirectional eddy currents in the attractive screen (2) and the sheet-metal blank (3). If the screen (2) is rigidly mounted on an insulated plate of the tool, then, in accordance with Ampere's law, the sheet-metal blank (3) is attracted to the working surface of the inductor (1).

It is necessary to emphasize the mandatory condition required for the operation of inductor systems with accessory screens. According to Ampere law, the force of attraction between the sheet-metal blank and the screen with unidirectional currents is possible only in the regime of intensive penetration of electromagnetic fields through the blank as well as the screen if (1) the identical electromagnetic fields between them cancel each other out and (2) the efficiency of the attraction process is defined by the intensity of electromagnetic field penetrated through the sheet-metal blank. The schematic of applied forces is shown in Fig. 3.12.

To accomplish an efficient subtraction of intensities of the electromagnetic field generated by the currents induced in the sheet-metal blank and in the attractive screen in the area between the screen and the blank, the induced currents in the screen and in the blank should be identical in terms of frequency and amplitude. This can be accomplished by fabricating the screen from approximately the same material and of the same thickness as the sheet-metal blank.

However, it should be emphasized that the efficiency of the proposed process is unlikely to be comparable with the efficiency of traditional EMF processes for one

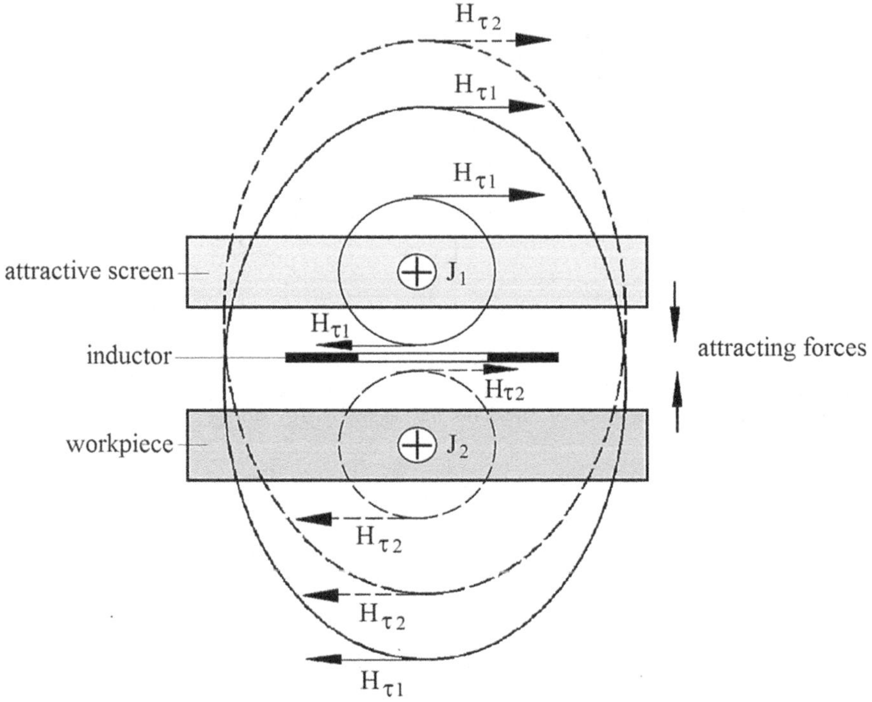

Fig. 3.12 Scheme of the physical processes in ISAS illustrating attracting forces between the screen and the blank. $J_{1,2}$—represents the unidirectional currents induced in the screen and in the blank and $H_{\tau 1,2}$—represents the tangential components of the magnetic-field intensity being excited by the currents $J_{1,2}$

very important reason: The electromagnetic field behind the blank and behind the screen are not limited with anything to spread in space, whereas in traditional EMF processes the electromagnetic field is contained within an equivalent clearance, which is the sum of the geometric clearance between the blank and the coil as well as the skin layers of both the blank and the coil. The pressure applied to the blank in traditional processes can be calculated as a ratio of energy in the coil divided by the volume of the equivalent clearance. In the proposed process, the equivalent clearance is much larger; therefore, the applied pressure is substantially lower. This can be considered a deficiency of the proposed process. However, the estimated cost of the energy required for one discharge of 10 kJ based on $0.09/kWh is 0.025 c. The cost of labor for 5-second operation based on $50/hour of fully burden cost is 7 c, which is 300 times higher. Therefore, from a purely economic perspective, the cost of electric energy is not very significant and is not expected to drive the cost of the manufacturing process.

3.2.2 Theoretical Analysis of EMF Attraction in ISAS

The interaction between a single-turn inductor, an attractive screen, and a sheet-metal blank was analyzed. To simplify the model and obtain a final solution in analytical form, a number of simplifying assumptions were made. Both the sheet-metal blank and the screen were considered to be identical thin sheets in order to have the most efficient subtraction of electromagnetic field in the area between the sheet-metal blank and the screen as was described in the previous section. The structural strength of the screen against the attracting forces was provided by attaching it to a massive dielectric plate. Both the sheet-metal blank and the screen were non-ferromagnetic metals with thickness d and specific conductivity $-\gamma$. They were located at the same distance—h from the single-turn inductor, which served as the source of the magnetic field.

A schematic corresponding to the described model of the process in a cylindrical coordinate system is shown in Fig. 3.13.

The following assumptions were made in order to simplify the mathematical model of the process:

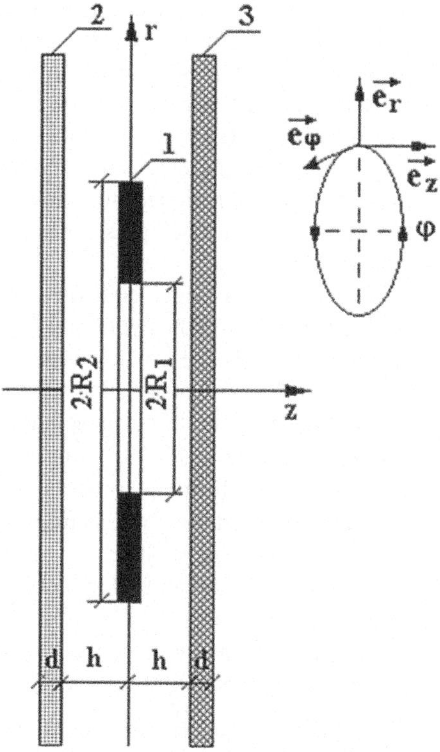

Fig. 3.13 Schematic of the single-turn inductor (1) between the two sheet-metal blanks (2 = the screen and 3 = the work-piece) where $\vec{e}_r, \vec{e}_\varphi, \vec{e}_z$ are the unit vectors in the cylindrical coordinate system

1. To simplify the analysis, the area where electromagnetic forces are being analyzed was subdivided into three zones: $r < R_1$; $R_1 < r < R_2$; and $r > R_2$. For the first and third zones, the effect of the inductor was not taken into account while analyzing the electromagnetic forces applied to the sheet-metal blank. For the second zone, the single-turn inductor was sufficiently thin (thickness $\rightarrow 0$) and "transparent" for the acting fields; thus, its metal had no effect on the considered electromagnetic processes. This assumption is justified by the low value of electrodynamic forces between the coil and the blank compared with the force between the blank and the screen.
2. The single-turn inductor had a flat cylindrical configuration with an internal radius R_1 and external radius R_2.
3. The system had an axial symmetry relative to the axis z; thus, $\partial/\partial\varphi = 0$ (where φ is the polar angle) and a plane symmetry relatively the plane of the single-turn inductor $(z = 0)$.
4. The sheet-metal blank, as well as the screen, was sufficiently thin, and its radial dimension was sufficiently large; thus, $d/R_{1,2} \ll 1$ and $\omega \cdot \tau \ll 1$ ($\tau = \mu_0 \cdot \gamma \cdot d^2$ where μ_0, γ and d are the vacuum permeability, specific conductivity, and thickness of the non-magnetic sheet-metal blank and $R_{1,2}$ indicates the internal and external radii of the inductor).
5. In the areas of the sheet-metal blank and the screen positioned against the single-turn inductor $r \in [R_1, R_2]$, only small repulsion forces exist, which can be explained by interaction of the oppositely directed currents in the inductor and in the sheet. In this case, the diffraction of electromagnetic field was neglected. A similar approach was often taken in the analysis of traditional EMF processes, for example, in the case when the tube was longer than the cylindrical coil in which it was positioned for radial compression on the mandrel. A typical assumption is that EMF pressure is applied only at the zone limited by the length of the coil.
6. The process was considered quasi-stationary according to the criterion $\omega/c \cdot \ell \ll 1$ where ω—is the cyclic frequency of the process; c is the light velocity for vacuum; and l is the greatest typical dimension of the system under consideration. Physically it means that the typical dimension of the system is substantially smaller than the wavelength of the acting electromagnetic field. For $l = 0.1$ m and $c = 3 \times 10^8$ m/sec, $\omega \ll 3.0 \times 10^9$ Hz, whereas for a typical electromagnetic-forming process, the cyclic frequency would not exceed 6×10^5 Hz, and for the electromagnetic-attraction process it is more than 1 order of magnitude smaller.
7. Reproduction of the mutual inductance of the tool coil and the sheet-metal blank was neglected.
8. Variability of conductivity as a consequence of temperature increase due to Joule heating was neglected.

The Maxwell equations in the L—space for the electromagnetic-field components being excited with zero initial conditions ($E_\varphi \neq 0$, $H_{r,z} \neq 0$) can be presented in their well-known form as follows:

$$\begin{cases} \dfrac{\partial H_r(p,r,z)}{\partial z} - \dfrac{\partial H_z(p,r,z)}{\partial r} = j_\varphi(p,r,z); \\ 1/r \cdot \partial/\partial r\big(r \cdot E_\varphi(p,r,z)\big) = -\mu_0 \cdot p \cdot H_z(p,r,z); \\ \dfrac{\partial E_\varphi(p,r,z)}{\partial z} = \mu_0 \cdot p \cdot H_r(p,r,z); \end{cases} \quad (3.13, 3.14, 3.15)$$

where p is the parameter of the Laplace transform;

$$E_\varphi(p,r,z) = L\{E_\varphi(t,r,z)\}; H_{r,z}(p,r,z) = L\{H_{r,z}(t,r,z)\};$$
$$j_\varphi(p,r\,z) = L\{j_\varphi(t,r,z)\}.$$

The current density in the right part of Eq. (3.13) is assumed as

$$j_\varphi(p,r,z) = (p \cdot \varepsilon_0 + \gamma) \cdot E_\varphi(p,r,z) + j_{\varphi i}(p,r,z), \quad (3.16)$$

where $j_{\varphi i}(t,r,z)$ is the current density in the inductor,

$$j_{\varphi i}(p,r,z) = j(p) \cdot F(r) \cdot \delta(z), \quad j(p) = \frac{I(p)}{R_2 - R_1}, I(p) = L\{I(t)\};$$

$F(r)$ is the function of radial distribution of current in the inductor turn, which can be assumed in the following form: $F(r) = \eta(r - R_1) - \eta(r - R_2)$;
$\delta\,(z)$ is the Delta-function; and
ε_0 is the dielectric permeability of vacuum.

To solve the formulated problem, the regions with homogeneous electrical physical characteristics must be selected. The geometrical and electrical dynamical symmetry of the explored system allow us to suppose that such regions include the space between the sheet blanks (non-conducting) ($z \in [-h, h]$), the identical sheets (the blank and the screen)—metal ($z \in [h, (h + d)]$), and the external system (non-conducting) ($z \in [d, +\infty)$). According to assumption 1, the single-turn inductor is sufficiently thin with its thickness $\to 0$ and "transparent" for the acting fields, so the metal of the inductor has no effect on considered electromagnetic processes. Therefore, there is no dedicated area for the inductor.

In each selected region, the system of differential Eqs. (3.13) through (3.15) must be integrated taking into account the form for the current density (3.16). According to the formulated assumptions, the equations for the azimuthal component of electrical-field intensity $E_\varphi(p,r,z)$ can be obtained.

In the space between the sheet blanks, $z \in [-h, h]$, we have

$$\frac{\partial^2 E\varphi(p,r,z)}{\partial z^2} + \frac{\partial}{\partial r}\left(\frac{1}{r} \cdot \frac{\partial}{\partial r}\left(r \cdot E_\varphi(p,r,z)\right)\right) \approx \mu_0 \cdot p \cdot j_{\varphi i}(p,r,z). \quad (3.17)$$

In the sheet metal (blanks and screen) $z \in [h, (h+d)]$:

$$\frac{\partial^2 E\varphi(p,r,z)}{\partial z^2} + \frac{\partial}{\partial r}\left(\frac{1}{r} \cdot \frac{\partial}{\partial r}(r \cdot E_\varphi(p,r,z))\right) - (p \cdot \gamma \cdot \mu_0) \cdot E_\varphi(p,r,z) = 0. \tag{3.18}$$

The external system, $z \in [(h+d), \infty)$:

$$\frac{\partial^2 E\varphi(p,r,z)}{\partial z^2} + \frac{\partial}{\partial r}\left(\frac{1}{r} \cdot \frac{\partial}{\partial r}(r \cdot E_\varphi(p,r,z))\right) \approx 0. \tag{3.19}$$

The axial symmetry of the system being considered and the requirement of the limit value for the electromagnetic field intensity for $r = 0$ and $r = \infty$ allow us to use the integral Fourier–Bessel transformation.

$$E_\varphi(p,r,z) = \int_0^\infty E_\varphi(p,\lambda,z) \cdot J_1(\lambda \cdot r)\lambda \mathrm{d}\lambda, \tag{3.20}$$

where $J_1(\lambda \cdot r)$ is the first-order Bessel function, and λ is the parameter of the Fourier–Bessel transformation.

In accordance with (3.20), Eqs. (3.17) through (3.19) are transformed to ordinary second-order differential equations:

(a) in the space between the metallic sheets (the blank and the screen), $z \in [-h, h]$:

$$\frac{\mathrm{d}^2 E_\varphi(p,\lambda,z)}{\mathrm{d}z^2} - \lambda^2 \cdot E_\varphi(p,\lambda,z) = K(p,\lambda) \cdot \delta(z), \tag{3.21}$$

where $K(p,\lambda) = \mu_0 \cdot p \cdot j(p) \cdot f(\lambda);\quad f(\lambda) = \int_0^\infty F(r) \cdot J_1(\lambda \cdot r) \cdot r \cdot \mathrm{d}r;$

(b) in the sheet-metal blank and the screen, $z \in [h, (h+d)]$:

$$\frac{\mathrm{d}^2 E_\varphi(p,\lambda,z)}{\mathrm{d}z^2} - q^2(p,\lambda) \cdot E_\varphi(p,\lambda,z) = 0, \tag{3.22}$$

where $q(p,\lambda) = \sqrt{\lambda^2 + p \cdot \mu_0 \cdot \gamma}$ is the propagation number in non-ferromagnetic metal with specific conductivity γ;

(c) external system, $z \in [(h+d), \infty)$:

$$\frac{\mathrm{d}^2 E_\varphi(p,\lambda,z)}{\mathrm{d}z^2} - \lambda^2 \cdot E_\varphi(p,\lambda,z) = 0. \tag{3.23}$$

General integrals of differential Eqs. (3.21) through (3.23) for the selected regions have the following form:

(a) in the space between the metal sheets, $z \in [-h, h]$, the function, satisfying to the symmetry condition relatively the plane $z = 0$, is:

$$E_{\varphi}^{(1)}(p,\lambda,z) = C(p,\lambda)\cdot \mathrm{ch}(\lambda z) + \frac{K(p,\lambda)}{\lambda}\cdot\left(\eta(z)\cdot \mathrm{sh}(\lambda z) + 0.5\cdot \mathrm{e}^{-\lambda z}\right), \tag{3.24}$$

where $C(p, l)$ is the integration constant;

(b) in the metal sheets, $z \in [h, (h+d)]$:

$$E_{\varphi}^{(2)}(p,\lambda,z) = D_1(p,\lambda)\cdot \mathrm{e}^{q(p,\lambda)\cdot(z-h)} + D_2(p,\lambda)\cdot \mathrm{e}^{-q(p,\lambda)\cdot(z-h)}, \tag{3.25}$$

where $D_{1,2}(p,\lambda)$ are the integration constants;

(c) external system, $z \in [(h+d),\infty)$, the function, satisfying the condition of limitation for $z \to \infty$, is:

$$E_{\varphi}^{(3)}(p,\lambda,z) = B(p,\lambda)\cdot \mathrm{e}^{-\lambda(z-(h+d))}, \tag{3.26}$$

where $B(p,\lambda)$ is the integration constant.

The tangential component of the intensity of the magnetic field can be found from Eq. (3.15) using Eqs. (3.24) through (3.26):

(a) $z \in [-h, h]$,

$$H_r^{(1)}(p,\lambda,z) = \frac{\lambda}{p\mu_0}\left(C(p,\ \lambda)\mathrm{sh}(\lambda z) + \frac{K(p,\lambda)}{\lambda}\left(\eta(z)\mathrm{ch}(\lambda(z) - 0.5\mathrm{e}^{-\lambda z})\right)\right), \tag{3.27}$$

(b) $z \in [h, (h+d)]$,

$$H_r^{(2)}(p,\lambda,z) = \frac{q(p,\lambda)}{p\mu_0}\left(D_1(p,\lambda)\mathrm{e}^{q(p,\lambda)(z-h)} - D_2(p,\lambda)\mathrm{e}^{-q(p,\lambda)(z-h)}\right), \tag{3.28}$$

(c) $z \in [(h+d),\infty)$:

$$H_r^{(3)}(p,\lambda,z) = -\frac{\lambda}{p\mu_0}\cdot B(p,\lambda)\cdot \mathrm{e}^{-\lambda(z-(h+d))}. \tag{3.29}$$

The systems of algebraic equations for defining the unknown constants of integration in general integrals (3.24) through (3.29) can be obtained from the continuity condition for the tangential components of the intensity of the electromagnetic field at the boundaries of the selected regions.

For $z = h$

$$\begin{cases} C(p,\lambda)\cdot \mathrm{ch}(\lambda h) + \dfrac{K(p,\lambda)}{2\lambda}\cdot e^{\lambda h} = D_1(p,\lambda) + D_2(p,\lambda); \\ C(p,\lambda)\cdot \mathrm{sh}(\lambda h) + \dfrac{K(p,\lambda)}{2\lambda}\cdot e^{\lambda h} = \dfrac{q(p,\lambda)}{\lambda}\cdot (D_1(p,\lambda) - D_2(p,\lambda)). \end{cases} \quad (3.30)$$

For $z = (h + d)$,

$$\begin{cases} D_1(p,\lambda)\cdot e^{q(p,\lambda)\cdot d} + D_2(p,\lambda)\cdot e^{-q(p,\lambda)\cdot d} = B(p,\lambda); \\ \dfrac{q(p,\lambda)}{\lambda}\cdot \left(D_1(p,\lambda)\cdot e^{q(p,\lambda)\cdot d} - D_2(p,\lambda)\cdot e^{-q(p,\lambda)\cdot d}\right) = -B(p,\lambda). \end{cases} \quad (3.31)$$

In the end, we are interested in the fields excited in the metal of the screen and sheet-metal blank. Eliminating $C(p,\lambda)$ and $B(p,\lambda)$ in the systems of the linear algebraic Eqs. (3.30) and (3.31), the unknown constants of integration $D_1(p,\lambda)$ and $D_2(p,\lambda)$ can be found.

Substituting in formulas (3.25) and (3.28) and Fourier–Bessel inversion, the formula for the intensities of the electromagnetic field can be written as dependence on the space variable connected with thickness of the sheet-metal blank ($\zeta \in [0, d]$).

$$E_{\varphi}^{(2)}(p,r,\zeta) = \frac{p\mu_0 j(p)}{2}\int_0^{\infty} \frac{f(\lambda)}{G(p,\lambda)} \times \left(\mathrm{sh}(q(p,\lambda)(\zeta - d)) - \frac{q(p,\lambda)}{\lambda}\mathrm{ch}(q(p,\lambda)(\zeta - d))\right)\cdot J_1(\lambda r)\mathrm{d}\lambda \quad (3.32)$$

where

$$G(p,\lambda) = \mathrm{sh}(q(p,\lambda)d)\cdot\left[\mathrm{sh}(\lambda h) + \left(\frac{q(p,\lambda)}{\lambda}\right)^2 \mathrm{ch}(\lambda h)\right] + \mathrm{ch}(q(p,\lambda)d)\cdot\frac{q(p,\lambda)}{\lambda}\cdot e^{\lambda h}$$

$$H_r^{(2)}(p,r,\zeta) = \frac{j(p)}{2}\cdot\int_0^{\infty} \frac{f(\lambda)\cdot q(p,\lambda)}{G(p,\lambda)} \times \left(\mathrm{ch}(q(p,\lambda)(\zeta - d)) - \frac{q(p,\lambda)}{\lambda}\mathrm{sh}(q(p,\lambda)(\zeta - d))\right)\cdot J_1(\lambda r)\,\mathrm{d}\lambda \quad (3.33)$$

To calculate the forces excited in the system, the currents induced in the metal screen and the sheet blank should be found.

The eddy currents density can be defined using (3.16) and (3.32):

$$j_{\varphi}(p,r,\zeta) = \frac{\mu_0\gamma(p\cdot j(p))}{2}\cdot\int\limits_0^{\infty}\frac{f(\lambda)}{G(p,\lambda)}\times\left(\mathrm{sh}(q(p,\lambda)(\zeta-d)) - \frac{q(p,\lambda)}{\lambda}\mathrm{ch}(q(p,\lambda)(\zeta-d))\right)\cdot J_1(\lambda r)\mathrm{d}\lambda \quad (3.34)$$

In the low-frequency regime when the intensive penetration of the inductor field takes place through the screen and the sheet-metal blank, the propagation number $q(p,\lambda)$ admits evaluation according to assumption 4 in the problem formulation:

$$|p\cdot\mu_0\cdot\gamma| \ll \lambda^2 \text{ and } q(p,\lambda)\approx\lambda. \quad (3.35)$$

Next, a limiting process should be fulfilled in (3.34) with condition (3.35).

After Laplace inversion, an analytical expression for density of the eddy currents induced in the metal of the screen and the sheet-metal blank can be obtained.

$$j_{\varphi}(t,\zeta,r)\big|_{q(p,\lambda)\approx\lambda}\approx -\frac{\mu_0\gamma}{2}\cdot\frac{\mathrm{d}j(t)}{\mathrm{d}t}\cdot\int\limits_0^{\infty} f(\lambda)\cdot e^{-\lambda\cdot(h+\zeta)}J_1(\lambda r)\cdot \mathrm{d}\lambda \quad (3.36)$$

As it follows from Eq. (3.36), the induced currents are proportional to the first time derivative of the inductor current. This result is verified by experimental waveforms in Fig. 3.2b and is in full accordance with the expression provided by Batygin et al. (2013).

The radial distribution of the induced currents can be defined from formula (3.36) by integration in limits $\zeta\in[0,d]$.

$$J_{\varphi}(t,r) = -\frac{\mu_0\gamma}{2}\cdot\frac{\mathrm{d}j(t)}{\mathrm{d}t}\cdot\int\limits_0^{\infty} f(\lambda)\cdot e^{-\lambda h}\cdot\frac{(1-e^{-\lambda d})}{\lambda}\cdot J_1(\lambda r)\,\mathrm{d}\lambda. \quad (3.37)$$

The surface density of attraction forces between the conductors located at a distance $(2h)$ is being defined based on the Ampere law as the value that is directly proportional to the product of the induced currents (3.37).

After identical transformation and introduction of symbols, which are convenient for numerical evaluation, the following expression can be obtained:

$$P_{\mathrm{attr}}(t,r) = \frac{\mu_0}{2\pi}\cdot\frac{J_{\varphi}^2(t,r)\cdot(2\pi\cdot r)}{(2h)} = \frac{\mu_0(\omega\cdot\tau)^2}{8h}\left(\frac{\mathrm{d}j(\phi)}{\mathrm{d}\phi}\right)^2\left(\int\limits_0^{\infty} F(\lambda,R_{1,2})\cdot e^{-\lambda h}\frac{(1-e^{-\lambda d})}{\lambda}J_1(\lambda r)\,\mathrm{d}\lambda\right)^2 \cdot r\times[\eta(r)-(\eta(r-R_1)-\eta(r-R_2))] \quad (3.38)$$

where $F(\lambda, R_{1,2}) = \frac{1}{(\lambda d)^2} \int\limits_{(\lambda \cdot R_1)}^{(\lambda \cdot R_2)} F(x) \cdot x \cdot J_1(x) \cdot \mathrm{d}x$, $\phi = \omega \cdot t$, is the phase, and ω is the cyclic frequency of the exciting current.

The integral of expression (3.38) on the circle with radius R_1 gives the formula for calculations of the integral force attracting the sheet blank to the screen in the internal round opening of the single-turn inductor.

$$
\begin{aligned}
F_{\mathrm{attr}}(t, r) &= \frac{\mu_0}{2\pi} \cdot \frac{2\pi \cdot \int\limits_0^{R_1} J_\varphi^2(t, r) \cdot (2\pi \cdot r) \cdot r\,\mathrm{d}r}{(2h)} \\
&= \frac{\mu_0 \pi \cdot (\omega \cdot \tau)^2}{(4 \cdot h)} \left(\frac{\mathrm{d}j(\phi)}{\mathrm{d}\phi} \right)^2 \\
&\quad \cdot \int\limits_0^{R_1} \left(\int\limits_0^{\infty} F(\lambda, R_{1,2}) \cdot \mathrm{e}^{-\lambda h} \frac{(1 - \mathrm{e}^{-\lambda d})}{\lambda} J_1(\lambda r)\,\mathrm{d}\lambda \right)^2 \cdot r^2 \mathrm{d}r
\end{aligned}
\tag{3.39}
$$

The analysis of the received results leads to the following conclusions.

As it follows from expressions (3.38) and (3.39), the time dependence of the attraction force is defined by square of the inductor-current derivative. This fact indicates constancy in time as well as constant direction of the excited forces.

At the same time, the calculation of the impulse of force shows that $S_{\mathrm{attr}} = \int\limits_0^{\infty} P_{\mathrm{attr}}(t, r)\,\mathrm{d}t \sim \int\limits_0^{\infty} \left(\frac{\mathrm{d}_{j(t)}}{\mathrm{d}_t} \right)^2 \mathrm{d}t > 0$, which confirms formally a cumulative character of deformations during the time of impulse of force action.

In the considered low-frequency regime of the acting fields, the attraction-field amplitude is directly proportional to the square of the working frequency and the specific conductivity of the screen and the sheet-metal blank. These forces are directed outside the area of the single-turn inductor where $r \in [0, R_1) \cup (R_2, \infty)$. According to assumption 5, in the area of the single-turn inductor where $r \in [R_1, R_2]$, only repulsion forces should exist, which can be explained by the interaction of oppositely directed currents in the inductor and in the sheet. This assumption leads to an abrupt change in the pressure applied to the blank from intensive attraction to rather small repulsion. Obviously, this is a simplification that is dictated by the selected analytical method in which the area of the blank is subdivided into three zones: (1) $0 < r < R_1$, (2) $R_1 < r < R_2$, and (3) $r > R_2$. This approach neglects the diffraction of the electromagnetic field around the corner of the coil.

Using expression (3.33) and condition (3.35), the space-time dependence for the tangential component of the magnetic field can be defined as follows:

$$H_r(r,\zeta,t) \approx \frac{j(t)}{2} \cdot \int_0^{\infty} f(\lambda) \cdot e^{-\lambda \cdot (h+\zeta)} \cdot \lambda \cdot J_1(\lambda \cdot r)\,d\lambda \tag{3.40}$$

where $\zeta \in [0, d]$.

The magnetic pressure on the screen and the sheet-metal blank can be defined as an integral (the integration variable is $\zeta \in [0,\ d]$) of the vector product of the current density (3.36) and the radial component of the magnetic field intensity (3.40) multiplied by the vacuum permeability.

$$\begin{aligned} P_{\text{rep}}(r,t) = \mu_0 \cdot \frac{(\omega \cdot \tau)}{(4 \cdot d^2)} \cdot j(\phi) \cdot \frac{dj(\phi)}{d\phi} \\ \times \int_0^{\infty} \int_0^{\infty} f(\lambda_1) \cdot f(\lambda_2) \cdot e^{-(\lambda_1+\lambda_2) \cdot h} \cdot \frac{1 - e^{-(\lambda_1+\lambda_2) \cdot d}}{(\lambda_1+\lambda_2)} \\ \cdot \lambda_1 \cdot J_1(\lambda_1 r) \cdot J_1(\lambda_2 r)\,d\lambda_1 d\lambda_2 \end{aligned} \tag{3.41}$$

Analysis of expression (3.41) leads to the following conclusions.

In the considered low-frequency regime of the acting fields, the time function of the electromagnetic force is being defined by product of electric current in the inductor and its first derivative by time. Based on this consideration, the electromagnetic force between the coil and the blank is an oscillating function of time, which means that this force changes direction during the process of electromagnetic attraction between the blank and the screen. The force impulse can be calculated as an integral by time: $S_{\text{rep}} = \int_0^{\infty} P_{\text{rep}}(t,r)\,dt \sim \int_0^{\infty} \left(j(t) \cdot \frac{dj(t)}{dt}\right) dt = \frac{j^2(t)}{2}\Big|_0^{\infty} = \frac{j(\infty) - j(0)}{2} = 0$ which means the absence of interaction or sufficiently small repulsion caused by Lorentz forces.

3.2.3 Numerical Estimates

Numerical calculations were performed for an experimental inductor system where the power supply of the electromagnetic-forming machine had capacitance $C = 1000\,\mu\text{F}$, maximum voltage $U = 5\,\text{kV}$, and maximum stored energy $W = 12.5\,\text{kJ}$ and where single-turn inductors with different internal and external radii R_1 and R_2 were employed. The screen and the sheet-metal blank are identical thin-walled sheets with thickness $d = 0.001$ m from the non-magnetic metal with specific conductivity $\gamma = 0.2 \times 10^7 (1/\text{Om m})$. They are located symmetrically relative to the plane of the single-turn inductor at a distance of $h = 0.001 \div 0.003$ m from each other. In the conducted experiments, the time function of the inductor current density has the shape of an exponentially relaxing sine that is $j(t) = j_m \cdot e^{-\delta_0 \cdot \phi}$.

$\sin \phi$, where $j_m = I_m/(R_2 - R_1)$ is the amplitude of the current density; I_m is the current amplitude; δ_0 is the relative decrement of relaxation; $\varphi = \omega \cdot t$ is the phase; and ω is the cyclic frequency. The electric-current amplitude in the inductor was $I_m = 40$ kA, and the frequency in the pulse was $f = 2$ kHz. The relative damping coefficient was calculated as the damping coefficient divided by the cyclic frequency and was equal to $\delta_0 = 0.3$. The current distribution in the inductor coil along the radial coordinate was approximated by the following function: $F(r) = \eta(r - R_1) - \eta(r - R_2)$.

The most illustrative results of the calculations fulfilled with the help of formulas (3.38) and (3.39) are represented below. In Figs. 3.14 and 3.17, P_{attr}^{max} are the maximum values of the distributed attraction forces were defined for $\phi_{max} \approx 0.78$ rad. In Fig. 3.15, F_{attr}^{max}, is the maximum value of the attraction force that was obtained by integrating the distributed P_{attr}^{max} along the internal area of the coil with the radius R_1.

The effect of the radius R_1 on the attraction pressure can be understood by considering the amount of energy in the single-turn coil defined by the formula $E = 0.5\, L{\cdot}I^2$. In this formula, the inductance of the coil can be considered as a linear function of the radius R_1. The larger the coil, the more energy is introduced into the system to support the maximum current $I_m = 40$ kA at the same frequency of the discharge $f = 2$ kHz. The illustrative graphs in Fig. 3.6 are based on the assumption that the frequency of the discharge does not depend on the inductance of the coil. In EMF-attraction processes, the pulsed transformer is expected to be employed to improve the efficiency of the single-turn coil system. The inductance of the pulsed transformer will dominate and make the frequency of the discharge consistent with different diameters of the single-turn coil.

The effect of the width of the single-turn coil can be understood from the exciting current–density perspective: For the same value of the exciting current $I_m = 40$ kA, the current density in the coil is greater if the width of the turn is smaller. The current density in the coil will reflect in the current densities in the blank as well as the screen and leads to increased attractive pressure.

The effect of sheet-material conductivity requires special attention. Different process parameters are required if Al alloy—compared with stainless steel—is subjected to electromagnetic attraction. According to formula (3.39), the attraction force is proportional in the square of the product: $\omega \cdot \tau = 2\pi \cdot f \cdot \gamma \cdot \mu_0$ where f, γ, μ_0—is frequency and electrical conductivity of the sheet metal and magnetic permeability of the vacuum. In order to obtain an identical attraction force, the following product should be the same: $(f \cdot \gamma) = \text{const}$. In other words, the frequency of discharge should be decreased in inverse proportion to the electrical conductivity in order to achieve the same attraction force.

A decrease of the frequency f to accommodate the needs of materials with better electric conductivity can be accomplished by increasing the inductance or capacitance of the system. If the capacitance of the system cannot be changed, then an efficient way to adjust the system is to employ the pulsed transformer. In traditional EMF processes, this element is also necessary to make the process efficient if a

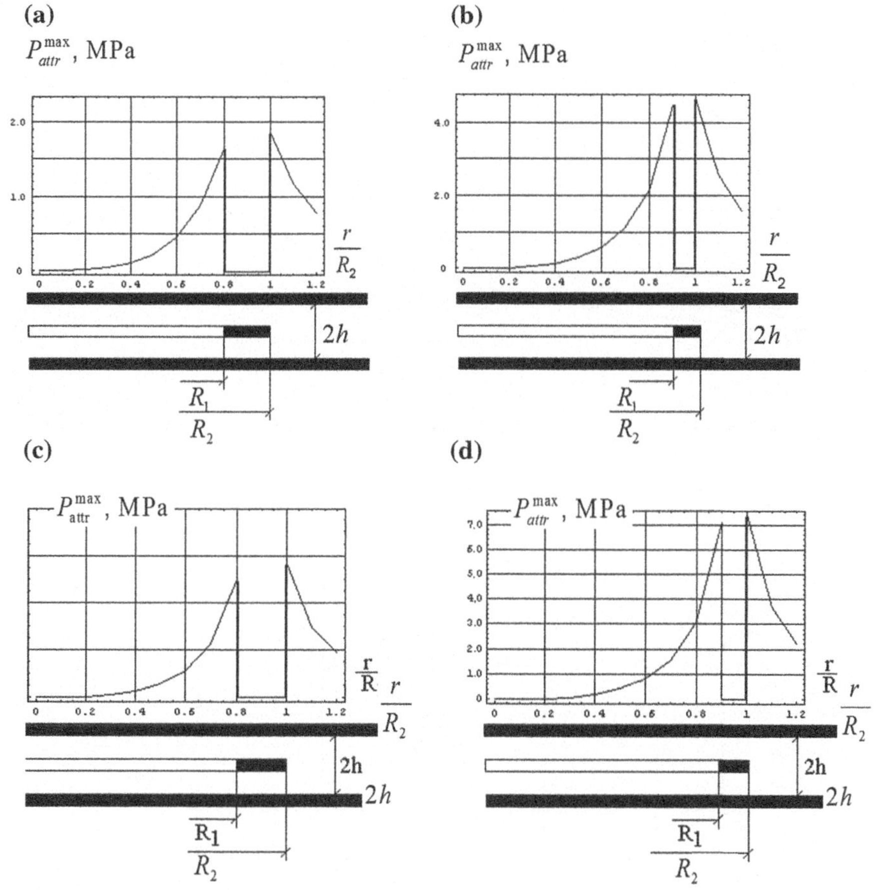

Fig. 3.14 Density distribution of attraction forces at the time moment when they attain maximum values for the following combinations of the dimensions of the inductor. (Top left panel) $R_1 = 0.015\,\text{m}, R_2 = 1/0.8 \cdot R_1$. (Top right panel) $R_1 = 0.015\,\text{m}, R_2 = 1/0.9 \cdot R_1$. (Bottom left panel) $R_1 = 0.035\,\text{m}, R_2 = 1/0.8 \cdot R_1$. (Bottom right panel) $R_1 = 0.035\,\text{m}, R_2 = 1/0.9 \cdot R_1$

small-diameter single-turn coil is employed for equipment with charging voltage of 5 kV where the inductance of the machine is several hundred nH. In the current laboratory set-up, no pulsed transformer was employed that was reflected in low efficiency: Rather high energy was required for a small displacement of the blank.

The main results of the fulfilled theoretical analysis can be formulated in the following way:

1. The highest concentration of the distributed forces of attraction occurs near the edges of the single-turn inductor (Fig. 3.14).
2. In the suggested ISAS, substantial amplitudes of the attraction forces are achievable: Their densities can reach approximately 7.0...8.0 MPa (Fig. 3.17d),

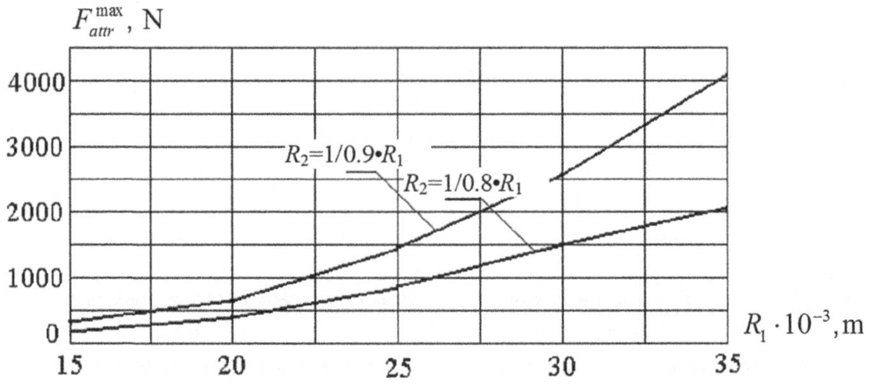

Fig. 3.15 Effect of the inner radius of the inductor on the attracting force between the screen and the blank for consistent distance between the screen and the sheet-metal blank $2h = 0.002\,\mathrm{m}$

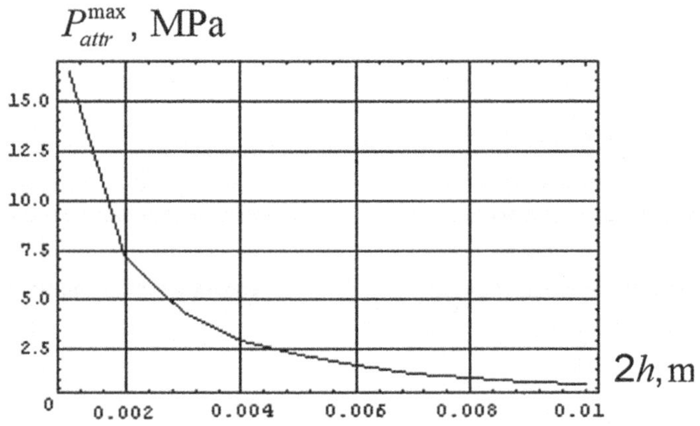

Fig. 3.16 Effect of the distance between the screen and the sheet-metal blank on the maximum value of the attraction-force density for $R_1 = 0.035\,\mathrm{m}$, $R_2 = 1/0.9 \cdot R_1$

and their integral values can reach approximately 4000 N (Fig. 3.15) on the circle square $\sim 0.001225 \cdot \pi\,\mathrm{m}^2$, accordingly. Attraction forces increase in square with the growth of the discharge electric current propagating through the coil. Decrease of the width of the coil for the same discharge current leads to an increase of attraction due to the increased density of the exciting current.

3. The amplitudes of the attraction forces depend essentially on the dimensions of the inner opening of the single-turn inductor, its width, and the distance between the screen and the sheet-metal blank (Figs. 3.15 and 3.16).
4. The increase of attraction forces with the growth of the radius R_1 can be explained by an overall increase of the area with induced currents in the screen

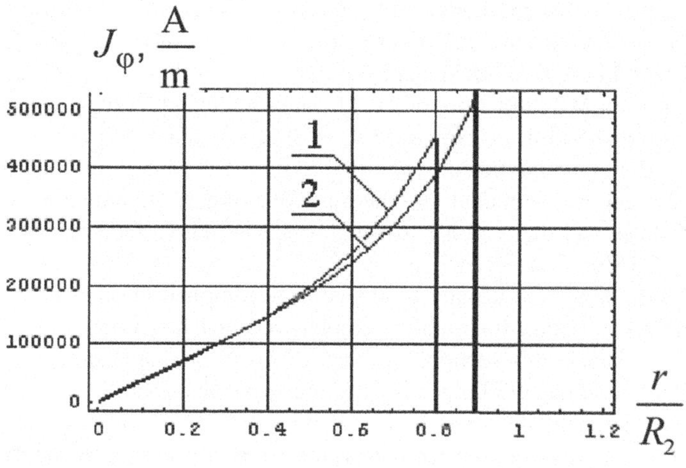

Fig. 3.17 Effect of the radius of the single-turn inductor on the density of the induced currents ($r \leq R_1 = 0.025$ m): (1) $R_1 = 0.8 \cdot R_2$ and (2) $R_1 = 0.9 \cdot R_2$

and in the blank. It can be also explained from an energy perspective: An increase of the radius of the coil leads to a linear increase of the inductance of the coil, which provides more energy into the coil for the consistent current amplitude and discharge frequency. The effect of width of the coil for consistent electric current can be explained by the current density increasing with a decrease of the width of the coil (Fig. 3.17).

5. As in traditional EMF processes where a decrease of the distance between the coil and the blank increases the pressure applied to the blank, a decrease of the distance between the screen and sheet-metal blank in ISAS results in increased electromagnetic-attraction force (Fig. 3.16).

3.2.4 Experimental Validation

The laboratory version of ISAS is shown in Fig. 3.18a. Structurally, the thin flat screen was fabricated from a 0.001 m–thick sheet of stainless steel that had the following chemical composition: 0.1% carbon, 9% nickel, and 18% chromium; the rest of the contents is iron. The material had a yield stress of 310 MPa and a specific conductivity of 0.2×10^7 1/Om m. The sheet-metal screen was rigidly mounted on the surface of a massive dielectric plate, thus preventing deformation of the screen. A flat single-turn inductor (thickness 0.001 m, internal diameter 0.08 m, and external diameter 0.1 m) was mounted on the insulation plate on top of the screen and covered by a layer of high-voltage insulation (0.0001-m thick) and clamped to the insulated surface of the screen. The blank was made from the same sheet as the screen (thickness 0.001 m, fabricated from stainless steel). The working zone of the investigated inductor system was the round inner cavity of its opening. The part of the sheet-metal blank positioned against the opening of the coil was expected to be deformed by the EMF-attraction forces. The amplitude and frequency of the electric current running through the coil were $J_m = 39.2$ kA and $f = 2$ kHz, respectively. The relative damping coefficient was equal to $\delta_0 = 0.3$. After eight discharges, the sheet-metal blank was attracted into the inner opening of the inductor: A dent with diameter 0.08 m and depth 0.001 m was formed on the surface of the sheet-metal blank (Fig. 3.18b).

Several discharges are usually produced to accomplish the desired deformation of the sheet-metal blank. It should be emphasized that for simple bulging of a flat sheet, as described in this section, two factors affect the displacement of the blank as a result of a single discharge: (1) the attractive distributed force required for plastic deformation of the sheet increases with a decrease of the radius from an infinite for the flat sheet to a certain value of R according to the formulation provided by Johnson and Mellor (1962); and (2) given the increased displacement from the blank being bulged, it comes much closer to the screen, and the attraction force increases substantially as can be seen in Fig. 3.16. In the experiments described previously, the second factor clearly dominates: During the first discharge, displacement of the blank is approximately 0.02 mm, whereas during the last discharge it is approximately 0.37 mm.

Assuming that the bulge can be approximated as a spherical arc, the equivalent quasi-static pressure that would form such a bulge can be calculated using the well-known analytical formulas for a circular diaphragm. The radius of the spherical arc was calculated as $R = \left(1 + (h/a)^2\right) \cdot a^2/2h$, where $a = 0.04$ m is half of the diameter of the bulged circular diaphragm, and $h = 0.001$ m is maximum height of the bulge. Based on these calculations, it was defined that $R = 0.8$ m. The quasi-static uniform bulge pressure was estimated for a flexible circular diaphragm

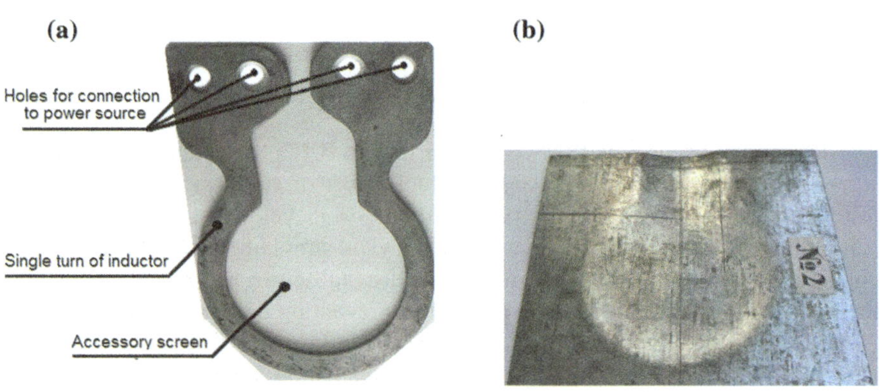

Fig. 3.18 Laboratory version of ISAS (**a**) and result of the EMF attraction of part of the stainless-steel sheet (**b**)

as $p = 2 \cdot YS \cdot t/R$ where YS is the yield stress of 310 MPa; $t = 0.001$ m; and $R = 0.8$ m. The quasi-static attracting pressure can be estimated as $P = 0.775$ MPa. The estimated quasi-static pressure was multiplied by the area of the working zone of 5024 mm^2. As a result, the maximum value of the attraction force was calculated as 3894 N.

The estimations of forces excited in tested ISAS were based on Eq. (3.27) and used the following parameters: the dimensions of the inductor were $R_1 = 0.004$ m, $R_2 = 0.005$ m; (3) the initial distance between the surface of the screen and surface of the sheet blank was 2 h = 0.0016 m.

According to the schematic shown in Fig. 3.13, the distance between the screen and the sheet-metal blank decreased from the initial 2 h = 0.0016 m to a final 2 h = 0.0006 m during the course of eight identical discharges. During the experimental work, the following measurements of maximum cumulative displacements were performed after each discharge: 1 = 0.02 mm, 2 = 0.06 mm, 3 = 0.1 mm, 4 = 0.16 mm, 5 = 0.25 mm, 6 = 0.4 mm, 7 = 0.63 mm; and 8 = 1.0 mm. The measurements were performed using a micrometer measure. However, the accuracy of such measurements should be processed as qualitative results providing a certain trend for maximum displacement of the formed bulge.

The distribution of density of the attractive forces between the blank and the screen was calculated using formula (3.39) in the working zone ($r \leq R_1 = 0.8 \cdot R_2$) for 2 h = 0.0016 m. The results of the conducted calculations are illustrated in Fig. 3.18a. Because the attractive forces were calculated for the flat blank, a certain averaging should be performed to account for the shape of the blank corresponding to a spherical segment. Because more intensive attractive forces are applied near the periphery of the coil, the average value correlating to approximately two thirds of the maximum displacement should be selected to correlate the attraction forces applied to the spherical segment and the flat sheet-metal blank for which all the analysis in this paper was performed. In this case, the distance between the blank and the screen is 0.93 mm, and the maximum attractive force at the end of the process can be calculated as 4850 N. The quasi-static force necessary to form the blank was 3894 N. Having the maximum attraction force exceeding the necessary quasi-static limit indicated that the electromagnetic force is capable of further forming the blank if the coil and screen are gradually moved away from the sheet-metal blank (Fig. 3.19).

In general, the conducted experiments showed that ISAS with a single-turn circular inductor and two non-magnetic metal sheets, one of which is the accessory attracting screen and the second the sheet-metal blank, allows us to develop mechanical forces that are sufficient to create plastic deformations based on EMF-attraction forces.

Conclusions

1. The conducted theoretical analysis and experiments confirmed the proposed concept of the inductor system with an accessory attracting screen as the tool for EMF attraction of the non-magnetic metals in the low-frequency range of electromagnetic fields when repulsion caused by the Lorentz forces is negligible

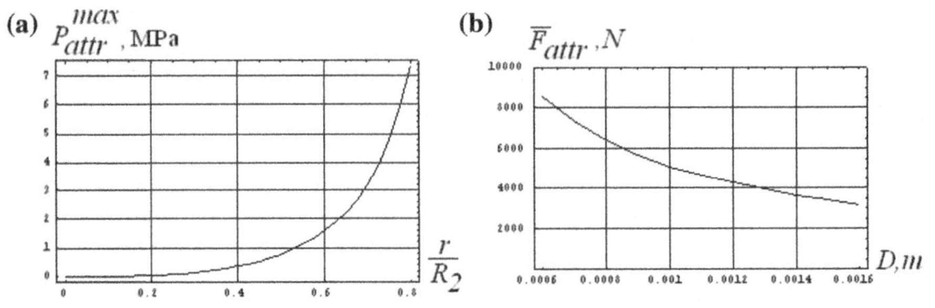

Fig. 3.19 Calculated space dependencies of the attracting efforts being excited for the tested ISAS. **a** Radial distribution of the attraction force-density maximum in the working zone, $r \le R_1 = 0.8 \cdot R_2$. **b** Dependence of the attracting force-square average on the depth of the bulge being formed

in comparison with the attraction forces of conductors with the unidirectional induced currents. The developed analytical model provides an estimation of attracting forces applied to the sheet-metal blank for specific parameters of the electric discharge.

2. The fulfilled experiments confirmed the theoretical conclusions and indicated that inductor systems with attracting screens provide sufficient attractive pressure to deform materials such as a 0.001 m-thick stainless-steel sheet.

3.3 Some Possible Applications of the EMF-Attraction Process and Main Conclusions

The proposed method of EMF attraction for ferromagnetic or non-ferromagnetic metal blanks is expected to serve as an initial concept for future development of effective tools designs for specific practical applications. Several possible applications listed below should be viewed as possible directions of the future work that may decrease cost and provide unique opportunities for single-sided forming and repair operations.

1. The EMF-attraction process can be employed for dent-repair procedures as was pointed out previously. Any versions of the EMF-attraction processes described previously substantially simplify the requirements of the EMF equipment: Two frequencies of interrupted discharge are no longer necessary. Instead they require a slow single-frequency low-voltage electric discharge, which enables the use of cheaper and longer-life capacitors, which substantially decreases safety issues and also extends the life of the coil insulation.
2. The EMF-attraction process with the ISAS tooling design can be employed for dent-repair procedures for aluminum panels as well as other non-magnetic materials including some galvanized panels. As it was described in the literature

sources including Batygin et al. (2013), electromagnetic-attraction technology for ferromagnetic and non-ferromagnetic metals has potential to provide shape restorations without damaging the paint and protective coating and without the necessity of having access to both sides of the panel to conduct the repair. For practical use of the suggested method with the help of ISAS, a number of coils assembled together with screens can be used to repair specific dents. To repair each individual dent, coils of several sizes can be employed sequentially. For example, initially a smaller coil can be employed to start pulling the sheet material from the bottom of the dent if the depth of the dent is large enough to make the coil covering the whole dent inefficient due to the distance between the surface of the blank and the surface of the coil along the vertical coordinate.

3. EMF-attraction processes may offer additional capabilities to traditional sheet metal–forming operations. Attempts to distribute the strains in sheet-metal blanks more uniformly during stamping operations often require the introduction of a third action of the press or an additional pre-forming operation as has been described by Golovashchenko et al. (2011). The pre-forming operation can be combined with the drawing process by positioning the attraction EMF coil inside the punch and stretching a portion of the blank, which usually does not undergo any plastic deformation. This approach can decrease the level of maximum strains in the stamped blank. To achieve the depth of drawing larger than several thicknesses of sheet metal, the tool incorporating the coil and the screen, insulated from each other, can be gradually moved in the direction of drawing while a sheet material is drawn into the cavity as shown in Fig. 3.20. A similar approach in traditional EMF was taken by Vohnout et al. (2004) by positioning EMF coils on the radii of the punch and running multiple discharges to advance the blank.
4. EMF-attraction processes can also offer the capability of personalizing of mass-produced products. For example, custom-feature lines, shapes, or imprints can be formed on mass-produced cars and trucks by taking advantage of the EMF-attraction process as a truly single sided-forming process. A substantial advantage of this process is that it does not damage the paint on any class A surface of the vehicle. The schematic example of a possible tool configuration is shown in Fig. 3.21. The shape of the tool needs must be fabricated from the same non-conductive compound often used for the prototype dies. Because the shape is expected to be individual, an extended life of the tool is not required. To achieve greater attracting pressure, the energy of the process must be increased. The attracting pressure is expected to increase proportionally to the energy of discharge and in second power to the discharge current. To increase the attraction pressure to the required level of filling the individualized shape, the discharge current might need to be increased to the level of 150 kA, which should provide a peak pressure 14 times greater than shown in Fig. 3.16. Peak pressures up to 100 MPa can be anticipated. Positioning the coil and screen closer to the surface of the blank and covering the coil and screen with high-voltage insulation before casting the coil in a high-durometer polyurethane compound may enable further improvement of the pressure amplitude.

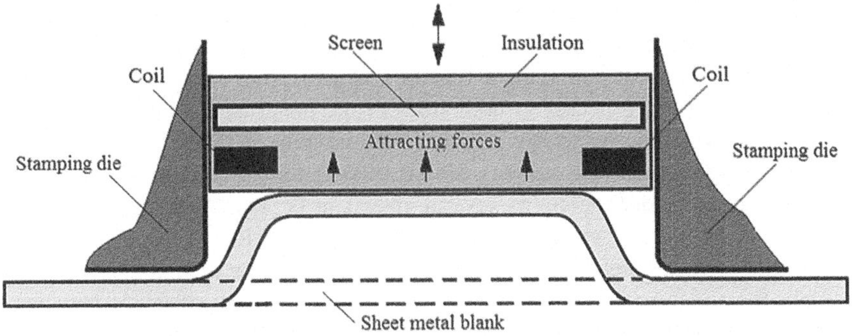

Fig. 3.20 Schematic drawing of the central portion of the blank into the punch cavity to stretch the area of the blank, which otherwise would be left with almost no deformations [described in detail by Golovashchenko et al. (2011, 2014)]

5. Additional opportunities can be created for forming materials at high temperatures where the flow stress of the material being formed is usually substantially decreased, whereas the electrical resistance is substantially increased. For low-frequency EMF processes, this is an almost perfect combination because increased electrical resistance leads to increased diffusion of the electromagnetic field. One of the issues that must be resolved to apply EMF processes for hot-forming processes is to develop high-voltage insulation materials that are capable of maintaining their insulation properties at high temperatures even though EMF-attraction processes do not require as high a voltage as do traditional EMF technologies.
6. Another possible application of EMF-attraction processes is in the applications where press-fit joints must be disassembled. In this case, the EMF-attraction approach must be modified to accommodate the tubular blanks. In this case, the attracting pulsed pressure of the proposed process must be superimposed with some relatively small axial quasi-static force, which would move the outer tube relative to the inner tube when the attractive pressure is applied. The schematic of such a process is shown in Fig. 3.22. The tooling incorporates two single-turn coils, and a tubular screen is positioned in between. If the inner tube is not

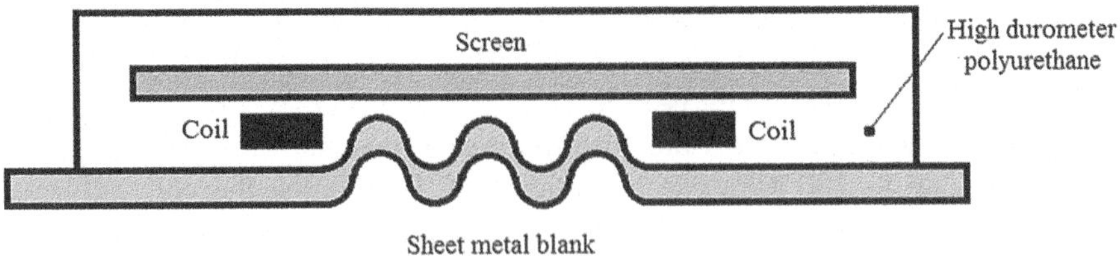

Fig. 3.21 Schematic of the tool to form an individual shape on a class-A surface

electrically conductive, the process will work similarly to the sheet attraction described earlier in the book. If the inner tube is conductive, different cases are possible: For example, the inner tube may undergo some elastic compression if the electromagnetic field does not penetrate entirely through the inner tube, whereas the outer tube is subject to electromagnetic attraction. This would lead to the disassembling of the joint.

The pressure required to disassemble the tubular joint can be estimated as $YS \times t/R$ to initiate the plastic deformation in the tube. For $YS = 180$ MPa for mild steels and for $t/R = 0.05$, the required quasi-static pressure is 9 MPa. Comparing this pressure level with the levels illustrated in Fig. 3.19a, it should be emphasized that a pulsed transformer must be employed to achieve an attraction pressure consistently greater than 9 MPa.

Concluding Remark

As it was indicated previously, the proposed schematics are conceptual illustrations of the potential use of the proposed EMF-attraction process.

Obviously, different potential applications described can be combined for some specific applications. One such possible areas is hot extrusion combined with EMF compression on a steel mandrel as described by Jaeger et al. (2011) where a hot EMF process is employed and where a potential issue is in the adhesion of tubes to internal mandrels. One of the defects of this highly efficient process is the sticking together of the aluminum tube and the steel mandrel. One possible method to address this issue can be the use of attraction pulsed electromagnetic force. A similar process, as described in possible application no. 5 and shown in Fig. 3.22, can be potentially employed in this case.

In general, the ISAS system may be applicable to ferromagnetic materials as well. However, the efficiency of such a system might be lower than the system described in Batygin et al. (2013). The advantage of using ISAS as a universal tool

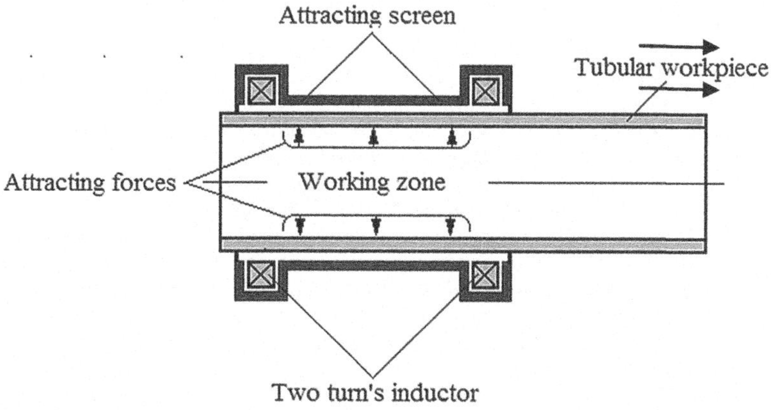

Fig. 3.22 Scheme of the process of disassembling of tubular joints using the EMF-attraction method

is technically very attractive. However, the necessity of fabricating an attractive screen for every sheet material-and-thickness combination may require additional expenses. In addition, the quantifiable comparison of efficiency of both systems requires additional analysis of the ISAS to be conducted taking into account additional magnetic forces.

Finally, future works will concentrate on the development of the tooling systems that would provide greater energy efficiency and more flexibility in applying attraction forces to the areas where they are the most required.

Relating to ISAS, moveable screens with moveable coils will invite special attention because they enable substantial expansion of the range of displacements achievable with the proposed process. In addition, a significant effort will be dedicated to the development of a high-efficiency pulsed transformer design, which would provide the necessary flexibility of working with different sizes of electromagnetic coils. An important step is to employ the numerical models capable of significantly decreasing the number of assumptions that were used in the analytical approach employed in the conducted consideration.

References

Batygin, Y. V., Chaplygin, E. A., Gnatov, A. V., & Golovashchenko, S. F. (2014). Pulsed electromagnetic attraction processes for sheet metal components. In *Proceedings of the 6th International Conference High Speed Forming 2014* (pp. 253–260), May 26–29, 2014, Daejeon, Korea.

Batygin, Y. V., Chaplygin, E. A., & Serikov, G. S. (2007). *Inductor for the angles magnetic pulsed forming in the bent sheet metal billets*. Patent of Ukraine 31309.

Batygin, Y. V., Gnatov, A. V., & Golovashchenko, S. F. (2013). Pulsed electromagnetic attraction of sheet metals—Fundamentals and specialized in perspective applications. *Journal of Materials Processing Technology, 213*(3), 444–452.

Batygin, Y. V., Gnatov, A. V., & Golovashchenko, S. F. (2014). Pulsed electromagnetic attraction of nonmagnetic sheet metals. *Journal of Materials Processing Technology,* 214(2), 390–401 (Elsevier).

Batygin, Y. V., Khavin, V. L., & Lavinskiy, V. I. (2006). *Method of magnetic pulsed attraction of thin-walled metal blanks*. Patent of Ukraine 74909.

Batygin, Y. V., Khimenko, L. T., & Lavinskiy, V. I. (2004). Direction change of the force action upon conductor under frequency variation of the acting magnetic field. In *Proceedings of the 1st International Conference on High Speed Forming*, Dortmund, pp. 157–160.

Bell, J. S., Kearsley, M. J., & Sykes, J. B. (1982). Revised and enlarged by E. M. Lifshitz & L. P. Pitaevskii. Oxford, New York, Beijing, Frankfurt: Pergamon Press, 620 p.

Belyy, I. V., Fertik, S. M., & Khimenko, L. T. (1977). *Spravochnik Po Magnitnoimpul'Snoy Obrabotke Metallov [Electromagnetic Metal Forming Handbook]*. English translation by Altynova, M. M., online available at http://www.mse.eng.ohiostate.edu/_Daehn/metalforminghb/index.html. Accessed 04.11.2010.

Daehn, G. S., Shang, J., & Vohnout, V. J. (2004). Improved formability by controlled of strain distribution in sheet stamping using electromagnetic impulses. In *Proceedings of the 1st International Conference on High Speed Forming* (pp. 211–220). Dortmund, Germany.

Iriondo, E., Gutiérreza, M.A., Gonzáleza, B., Alcarazb., & J.L., Daehnc G.S. (2011). Electromagnetic impulse calibration of high strength sheet metal structures. *Journal of Materials Processing Technology*. 211(5), 909–915.

Jaeger A., Risch D., & Tekkaya A.E. (2011). Thermo-mechanical processing of aluminum profiles by integrated electromagnetic compression subsequent to hot extrusion. *Journal of Materials Processing Technology*. 211, 936–943.

Johnson, W., & Mellor, P.B. (1962). *Plasticity for Mechanical Engineers*. London, D.Van Nostrand Company Ltd., pp.187–192.

Knoepfel, H. (1970). *Pulsed high magnetic fields* (pp. 34–36). Amsterdam and London: North-Holland Publishing Company.

Korn, A., & Korn. M. (1968). *Mathematical handbook for scientists and engineers*. Second enlarged and revised edition. New York, San Francisco, Toronto, London, Sydney: McGraw-Hill Company, 832 p.

Landau, L. D., Lifshitz, E. M. (1984). *Electrodynamics of continuous media* (pp. 105–113, 126–129, 199–210, v. 8 of Course of Theoretical Physics, Translated from Russian).

Serikov, G. S. (2010). *Research of electrodynamics processes in the inductor systems for the magnetic-impulsive molding of sheet metals [Russian]*. Manuscript. PhD thesis. Kharkov National Automobile and Highway University. 162 p. Kharkov 2010.

Shneerson, G. A. (1981). Polja i perekhodnye processy v apparature sverkhsilnykh tokov [Russian]. USSR, Leningrad: Publishing House "Energia.", 199 p.

Vohnout V.J., Shang J., Daehn G.S. (2004). Improved formability by controlled of strain distribution in sheet stamping using electromagnetic impulses. *Proceedings of the 1st International Conference on High Speed Forming*. Dortmund, Germany, pp.211–220.

Chapter 4
Practical Realization of External-Dent Removal in Car Bodies Using EMF Attraction

Application of EMF technologies in the automotive industry is an innovative technique for contactless surface treatment of sheet metals. This chapter describes the practical application of magnetic pulse systems for solving problems of contactless dent removal on surface-body elements without damaging paintwork. A detailed description of the test equipment, step-by-step execution of the dent-removal operation, and subsequent experiments with their results are presented. All experiments presented here were realized by equipment that had been developed as a prototype for future technologies. The most workable types of instruments employing different principles of dent removal were tested. The results expected showed possible advantages of the system developed compared with existing magnetic pulse systems. These advantages consist of the possibilities for non-magnetic metal processing as well as decreasing system size and its primary cost.

The present consideration is dedicated to the description of external dent-removal processes on the car bodies with help of the different tools of the magnetic pulsed attraction-type using low-frequency current discharges. The main features of these repair technologies are the following: (1) the working tool is placed on the vehicle body external surface; and (2) the dent removal can be realized without any disassembly and dismantling operations.

Tools of the represented technologies created on basis of the different principles of action were examined. The first of them suggests using the magnetic properties of the car steel panels. The second one is the recently developed tool design, ISAS, which uses the phenomenon of the mutual attraction of the uni-directed currents according to the Ampere law. The tool employing the first principle of action allows one to work with ferromagnetic metals only, whereas the ISAS tool is a universal tool giving the possibility to work with ferromagnetic as well as non-ferromagnetic metals.

Despite their different principle actions and the different appointments of the represented magnetic pulsed tools, the basic diagram and the algorithm of the repair

Y. Batygin et al., *Electromagnetic Metal Forming for Advanced Processing Technologies*, Materials Forming, Machining and Tribology,
https://doi.org/10.1007/978-3-319-74570-1_4

operations remain the same. They provide high effectiveness of a non-contact external flattening of a car body under real production conditions.

All represented works were performed in the Laboratory of the Electromagnetic Technologies of the Kharkov National Automobile and Highway University, Ukraine.

4.1 Experimental Approbation of the Process and Tool Used to Flatten Ferromagnetic-Steel Samples

The basic work diagram of the elaborated complex for the external magnetic pulsed removal of dents on a car body can be explained with help of the scheme shown in Fig. 4.1. The principle action of the tool for the repair operation being realized consists in exciting the forces of electrodynamical attraction at the expense of the magnetic properties of the sheet metal being processed. However, it should be noted once again that this scheme also can be applied for illustration of the repair process for non-magnetic metal as well.

The technical characteristics of the "power source" (in the special literature, "power source" is the term used for magnetic pulsed installation) are the following: the voltage of the electric network is U_0 = 220–380 V; the voltage of the charge of the capacitor bank is $U_C \approx 100 \div 2100$ V; the maximum of the stored energy is $W \approx 2$ kJ; the bank capacity is $C = 1200$ μF; the own inductance is $L = 440$ nH; the working frequency in the current pulse is $f_W = 7$ kHz; and the frequency of repeating the current pulses is $f_I \approx 1 \div 10$ Hz.

The tool of the electromagnetic attraction of the car's steel body is the inductor described in detail before Part 3.1. It is connected to the power source by an electrical cable. With help of this tool (it is laced on the external side of the car body), an operator accomplishes the dent removal on the automobile-panel sample. The work control of the complex is realized by devices of the "control system" on

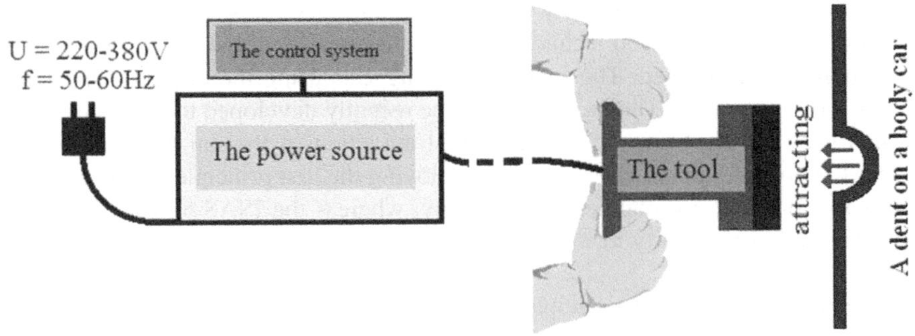

Fig. 4.1 Basic diagram of the elaborated complex for the external Electromagnetic pulsed removal of dents on a car body

the face panel of the "power source." A switch on the tool is provided for local control.

In detail, basically the work of the represented scheme is achieved in the following way. First, all data necessary for the successful fulfillment of the production operation are collected. The operator places the working surface of the tool over the dent on the car body and switches on the system. After repeating several pulses of the electromagnetic attraction forces, the dent will be removed. The operation is complete.

Let us pass to the practical algorithm of the external removal of the dents. We give more detailed information about the experimental equipment, some components of which are shown in Fig. 4.2. Basically, this equipment consists of two main functional components. The first of them is "the force block" (Fig. 4.2a). This is the aforementioned power source. Its technical characteristics were listed in the previous part of the present book. The second main functional component is the inductor, which is the tool used for the external flattening of the car body with help of the forces of electromagnetic attraction (Fig. 4.2c). The efficiency of the used tool is due to the essential influence of the magnetic properties of the dented steel with the low frequencies of the acting fields.

The algorithm as a sequence of the operations being fulfilled.

1. An external survey of the surface of a car body is fulfilled visually. The purpose is to determine and estimate the damage that requires repair. The shape and geometric dimensions of dents requires choosing a necessary level of attraction force as well as a value of necessary voltage of the capacitor bank.
2. The necessary voltage and quantity of the force actions are fixed by an operator on the remote controller.
3. A special dielectric insert is arranged over the dent on the surface of the body-panel sample. The aim consists of creating a hard working surface of the tool over the dent.

Fig. 4.2 Experimental complex for the electromagnetic flattening of the steel body of an automobile (**a**) the entire complex: 1 the used tool for flattening; 2 the cable connection; 3 the power source (MPIS 2 the magnetic pulsed installation on 2 kJ) (**b**) the remote controller (**c**) different tools for the electromagnetic attraction of dents in the sheet metal

4. The operator places the tool over the dent, switches on the voltage of the power source, and fulfills the external flattening the body panel sample of the car. Result: The excited forces attract the dent metal up to the level of the initial smooth surface. The proper completion of flattening is being visually.
5. The tool and insert are removed after completion of the force action. The dent has been successfully removed. The protective paint covering remains without any damage.

The represented algorithm of practical flattening is illustrated by photos in Fig. 4.3 where the dent removal on a body panel of the automobile "Audi" is shown.

Some Generalizations

1. The electromagnetic attraction process for deforming ferromagnetic sheet-metal materials, such as low-carbon steels, is suggested for the external non-contact removal of dents on car bodies.
2. The basic diagram of the elaborated complex for external electromagnetic flattening is represented.
3. The implementation of the external non-contact removal of dents on a car body is described as a practical algorithm of the sequence of the required operations.
4. The suggested flattening method allows preserving the paint covering without any damage, which was confirmed experimentally.

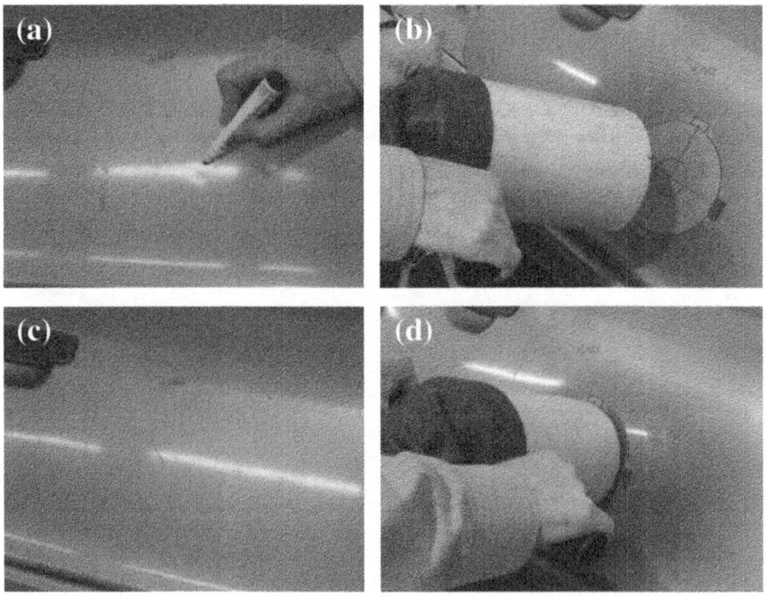

Fig. 4.3 Photographs representing of the electromagnetic flattening algorithm (**a**) fixation of a dent, which is the object of removal (**b**) centering the tool over the dent with help of a thin dielectric insert (**c**) realization of flattening (several applications of the attracting force) (**d**) door of the car after dent removal. The protective paint covering is preserved without any damage

4.2 Experimental Approbation of Magnetic Pulsed Dent Removal in Any Sheet Metal Using the Proposed Universal Tool

One should note some main statements made in the previous text of this book before reading the upcoming analysis of the experiments conducted.

The present exposition is dedicated to the experimental study of pulsed electromagnetic attraction with help of the recently developed tool design—ISAS—which is used to remove the external dents damaged car body. The concept of this system workability is based on mutual attraction of the induced uni-directed currents in the accessory screen and in the sheet-metal blank (Ampere law). Despite subsequent additional detailed information, it should be noted that the main feature of the last tool construction and deference from the previous option of the ISAS tool is that the multi-turn winding (i.e., exciting solenoid) is placed on the external side of the accessory screen. The currents induced in the flat accessory screen and sheet billet are excited by the penetrating magnetic field of the external source (the multi-turn winding, which is positioned behind the screen). Particularly it should be noted that this tool is capable of creating attraction forces of the sheet metal of any physical nature. The question that remains regards the magnetic conductors (the iron alloys) as well as non-magnetic metals (the stainless steel or aluminum).

Practical interest in the magnetic pulsed attraction of any metal is initiated by the growing demands on production operations for the effective restoration of damaged airplane and vehicle bodies. The repair process must be realized from the external side of the metal surface being restored. The dents must be removed without disassembling the bodies and without damage of the existing protective cover. The electromagnetic-forming (EMF) technologies satisfy all listed requirements.

Historically and chronologically, the USA can be considered a pioneer and leader in the field of technical-system creation for external dent removal in sheet conductors. The trustworthiness of this fact is confirmed by Web sites of the well-known Americans companies, Electroimpact and Fluxtronic, for example. They represent EMF equipment for the exterior restoration of aircrafts wings and fuselages when access from the inside of the panel is problematic if not impossible.

The European leader in area of the creation of modern repair equipment is the well-known company Betag Innovation. The company's Web site contains many suggestions of different modern technologies for flattening a car body. Among them there is the Magnetic Dent Remover. This device permits the elimination of dents by the attraction of designated parts of magnetic steel blanks only. Non-magnetic metals can not be restored. The undoubted advantage of the Magnetic Dent Remover is the possibility of restoration from the external side without disassembling the components and without damage of existing paint and protective coating. It should be noted that the principle action of this device was taken from Batygin et al. (2004). The physical essence of the mentioned principle action was justified theoretically and later experimentally. As was shown by Batygin et al. (2013) by the low working frequencies of the acting field, the attracting forces become prevailed over the repelling forces, which are the well-known forces of magnetic pressure. Attraction is caused by the magnetic properties of sheet billet metal. The nature of repulsion is caused by Lorentz forces.

The past ISAS constructions had an essential demerit: The winding of the exciting inductor was located between the screen and the work-piece. In connection with this, the attraction forces decreased, and the Lorentz repulsion was rather strong due to the proximity of the inductor and the work-piece sheet metal.

Further development of the theory and practice of the EMF-attracting tools based on Ampere's law have led to a new ISAS construction suggested by Batygin et al. (2013) where the exciting inductor is moved out of the working zone and positioned on the external side of the attracting screen. The new ISAS consists of the parallel lays of the flat winding of the exciting inductor, the accessory screen, and the sheet-metal billet. The induced eddy currents in the screen and the billet metal are excited by the penetrated pulse magnetic field generated by the field of the external exciting inductor.

A principle scheme of the experimental modeling of the production operation of non-contact dent removal by ISAS is represented in Fig. 4.4. Once again some features of the conducted experiments should be noted.

1. The external removal of the dent was realized from the side of concavity on the metal surface.
2. ISAS excitation was realized by the outside source of the magnetic field. The exciting-inductor winding was positioned on the external side of the accessory screen.

Equipment and the experimental samples.

The principle diagram of the ISAS experimental construction is represented in Fig. 4.4 (see a part termed "ISAS"). Its main components are the accessory attracting screen and the circular winding of the exciting inductor positioned on the external surface of the screen.

It should be noted that the proposed design of ISAS by the result of the action is similar to the well-known action of an electromagnet. however, unlike of last version ISAS permits to realize attraction of not only ferromagnetics but metals of any physical nature.

The ISAS inductor was connected to the power source. It was MPIS-2 (MAGNETIC PULSED INSTALLATION with 2 kJ of stored energy) created in the Laboratory of the Electromagnetic Technologies and described in detail by Batygin et al. (2013, 2014a, b).

The distinguished feature of the power source—MPIS-2—is its work in the regime as a generator of a designated repetition of a quantity of the current pulses. The current pulses under flowing in the inductor winding are exciting the according pulses of the attraction force of the sheet metal to the screen plane.

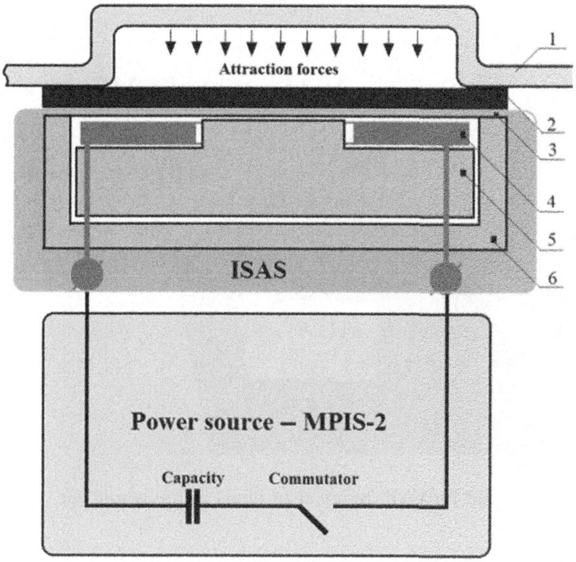

Fig. 4.4 Principle scheme of dent removal with the help of ISAS. 1 The sheet-metal blank; 2 (ISAS) the accessory attracting screen; 3 a dielectric insert, 4 the winding of the exciting inductor, 5 a dielectric compaction, 6 the dielectric body. Power source—MPIS-2 (magnetic pulse installation with 2 kJ of energy stored). Capacity—a capacitor storage; Commutator—a block of thyristor switches

The main data of the MPIS-2 are the following.

1. The stored energy equals to approbatory 2.4 kJ; the voltage is approximately 1500 V; and the current in pulse equals to approximately 12 кA.
2. The working frequency in discharge is approximately 1500 Hz.
3. The frequency repetition of the current pulses is approximately 5 Hz.
4. The quantity of the force attraction pulses that were necessary for the dent removing is approximately 18–20.

The general view of the experimental equipment, including the power source and the tool, is shown in Fig. 4.5.

The separate plates of the body sheet metals of automobiles "Subaru" and "Citroen" were taken as the experimental samples. Their dimensions were approximately 15 m $\times$ 0.15 m. The thickness of the samples was $\sim$0.0008 $\div$ 0.001 m. The depth of the dents which were created artificially was approximately 0.0018 $\div$ 0.002 m. Their external diameter was $\sim$0.05 m. Besides the separate sheet samples an object of the experiment was as well a door of the automobile "Audi" represented in Fig. 4.3a. The thickness of its steel covering was approximately 0.001 m. The dimensions of a natural dent were approximately $\varnothing$0.06 m $\times$ 0.02 m.

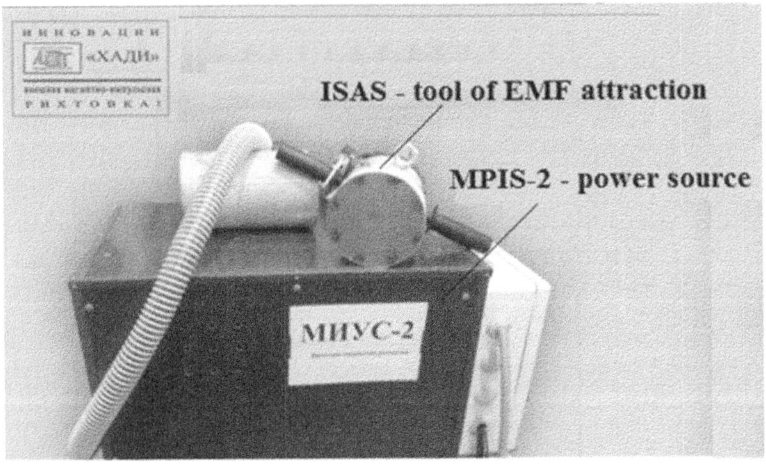

Fig. 4.5 Equipment needed for EMF removal of dents in the whole car body

Experimental Results

After switching on the mains, voltage was fed on the step-up transformer primary winding. Charging the capacitors was carried out through the rectifier up to a specified level of the stored energy. After this, the discharge circuit was closed automatically with help of a block of the thyristor commutators and the generated current pulses were fed in the winding of the exciting inductor (ISAS).

A cycle of charge–discharge and the attraction force pulses were repeated several times until the dent on the object surface of the experimental work = piece was removed. Some results of the external non-contact magnetic pulsed removal of the dents in a separate sheet samples of Japanese and European automobile coverings are shown in Fig. 4.6.

First it is necessary to note as an experimental positive result a level of force attraction that provided high effectiveness of the fulfilled production operation. Particularly it should be noted that flattening quality was not investigated in this experiment. The aim of the conducted experiment consisted in only a practical test of the ISAS as a tool of magnetic pulsed attraction.

The second positive result that the protective paint-and-lacquer coating was preserved. However, a quite important note remains: Saving the paint-and-lacquer or other protective coating depends on many factors that are not immediately connected precisely with the magnetic pulsed attraction of the deformed metal. Figure 4.7 illustrates a course of experiment of flattening the steel covering of an Audi automobile door with a dent.

The experiment of door restoration on an Audi automobile under real-world conditions demonstrated the effective fulfillment of the manufacturing operation. A visual control of the result shows the real possibility of qualitative EMF flattening with help of the suggested ISAS tool.

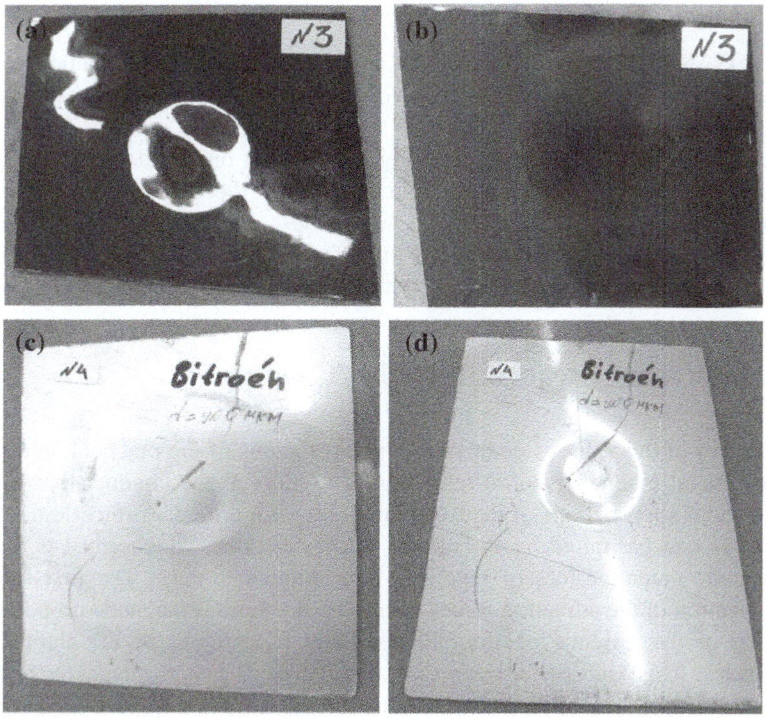

Fig. 4.6 Experimental samples of different automobile steels from a Subaru automobile—(**a**) before force action and (**b**) after attraction—and a Citroen automobile—(**c**) before force action and (**d**) after attraction

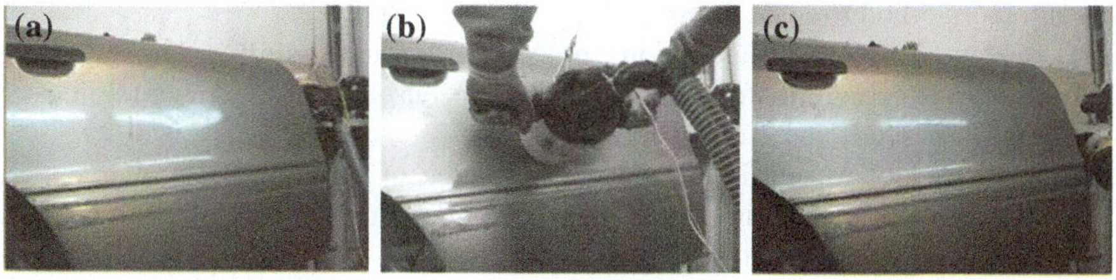

Fig. 4.7 External EMF flattening in the real repair process (**a**) the door before flattening (**b**) the ISAS tool in action (**c**) the door after flattening

Concluding Remarks

1. The ISAS tool, which is excited by an external source of magnetic field, was successfully approbated in the repair operation of external non-contact magnetic pulsed removal of dents.
2. Removing the dents was successfully realized not only under idealized conditions for separate steel samples but also under the real-world conditions of automobile-body repair.
3. The obtained results demonstrated the effectiveness of the suggested tool in a principally new repair technology with the use of pulsed magnetic-field energy.

4.3 Inferences Gleaned from this Chapter

The content of this book is dedicated to investigations of the practical applications of sheet-metal magnetic pulsed attraction. The aims of the conducted work were experimental approbations of the two main tools with different principles of action. Both tools can be united by the results of their actions, which consists in transforming the natural forces of repelling into attracting forces. The first tool uses intensification of the attraction forces caused by the metal's magnetic properties and damping of the Lorentz repelling forces under a decreases of the acting fields' frequencies. This kind of tool can only remove dents in ferromagnetic materials. The second tool is built on the introduction of a new constructive component termed the "accessory attracting screen," with the help of which it becomes possible to realize mutual attraction of low-frequency induced currents. The main results of the fulfilled experiments are generalized as follows.

1. Removing the dents was realized experimentally not only under idealized conditions for separate steel samples but also under the real-world conditions of automobile-body repair.
2. The suggested EMF-attraction process can successfully deform, in a low-frequency regime, ferromagnetic sheet metals such as the low-carbon steels found in modern automobile bodies.
3. The ISAS tool, which is excited by an source of magnetic field, was experimentally approbated in the repair operation consisting of the external non-contact magnetic pulsed removal of dents.
4. ISAS is a universal tool for dents removal from metals of any physical nature independently with and without magnetic properties.
5. In general, the obtained results demonstrated the effectiveness of the suggested tools in a principally new repair technology with use of pulsed magnetic-field energy.

As it indicated previously, the proposed schematics of the production operations are conceptual illustrations showing the potential use of the proposed EMF-attraction process.

Obviously, different potential applications described previously can be combined together for some specific applications. One such possible areas is hot extrusion combined with EMF compression on a steel mandrel where a hot EMF process is employed and where a potential issue is the adhesion of tubes to internal mandrels. One of the defects of this highly efficient process is the sticking that occurs between the aluminum tube and the steel mandrel. One of possible methods to address this issue can be magnetic pulsed attraction.

Returning to the universal ISAS tool, it should be noted that these systems may be applicable to ferromagnetic as well as non-ferromagnetic metals. The advantage of using ISAS as a universal tool is technically very attractive.

To finish this summation, all possible future works will concentrate on development of tooling systems that would provide greater energy efficiency and more flexibility in applying attraction forces to the areas where they are most required.

References

Batygin, Y. V. (2014). Experimental test of the tool for the external EMF removing dents on a car body. *International Journal of Energy and Power Engineering, 3*(4), 204–208. Online available at http://www.sciencepublishinggroup.com/j/ijepe. Published September 20, 2014.

Batygin, Y. V., Chaplygin, E. A., Gnatov A. V., & Golovashchenko, S. F. (2014a). Pulsed electromagnetic attraction processes for sheet metal components. In *Proceedings of the 6th International Conference on High Speed Forming 2014*, May 26–29, 2014, Daejeon, Korea, pp. 253–260.

Batygin, Y. V., & Gnatov, A. V. (2015). Basic diagram and practical algorithm removing dents on the body of vehicle by the pulsed electromagnetic attraction. *International Journal of Engineering Science and Management (IJESM), 5*(1), 47–52. Online available at http://www.ijesmjournal.com.

Batygin, Y. V., Gnatov, A. V., & Golovashchenko, S. F. (2013). Pulsed electromagnetic attraction of sheet metals—Fundamentals and specialized in perspective applications. *Journal of Materials Processing Technology, 213*(3), 444–452.

Batygin, Y. V., Gnatov, A. V., & Golovashchenko, S. F. (2014b). Pulsed electromagnetic attraction of nonmagnetic sheet metals. *Journal of Materials Processing Technology, 214*(2), 390–401.

Batygin, Y. V., Khimenko, L. T., & Lavinskiy, V. I. (2004). Direction change of the force action upon conductor under frequency variation of the acting magnetic field. In *Proceedings of the 1st International Conference on High Speed Forming*, Dortmund, pp. 157–160.

Electroimpact is a world leader in design and manufacturing of aerospace tooling and automation. Online available at www.electroimpact.com. Electromagnetic Dent Removal. Online available at www.fluxtronic.com (shown on 2014).

Chapter 5
Conclusion

The present book consists of three separate parts that illuminate two possible major applications of metal magnetic pulsed processing. This involves the stamping of angles in sheet metal using repelling forces in a regime when the magnetic-field penetration is practically absent (the skin effect) and the straightening of sheet metal using attraction forces in the regime of the acting field's intensive penetration.

According to the exposition order of the represented material, the generalization of the described investigations main results in the end-concluding view will be given below.

The first application is "angles filling" in sheet metals. The given term is a special one deriving from forging stamping production. In practical term, it designates the creation of sharp angles with rounding off, the level of which is demanded according to production-operation conditions.

Successful realization of this production operation is caused by a new property appearance in the behavior of the metal being worked under a high-speed magnetic pulsed action. This property, which was described by G. S. Dahen, Ph.D., is termed "hyper-plasticity." Its practical usage allows the stamping of materials using EMF methods without disassembling the parts or destroying protective the coating, unlike the known mechanical approaches.

Magnetic pulsed metalworking realizes practically a power force-acting regime in the time interval of a microsecond. The main particularity of these methods is the indissoluble connection of the inductor-tool with the billet, which is the subject of processing. During the mechanical stamping, if the efficiency is not connected directly with the inner processes in the processed metal, action of the magnetic-pulse force is physically possible only when an electromagnetic connection exists between the inductor-tool coil current and the current induced in the billet. Their interaction leads to the appearance of, high-power ponder-motor forces, the value of which is proportional to product of these currents' amplitudes.

The main results of Chap. 1 can be united by the following positions:

Y. Batygin et al., *Electromagnetic Metal Forming for Advanced Processing Technologies*, Materials Forming, Machining and Tribology,
https://doi.org/10.1007/978-3-319-74570-1_5

1. Analytical expressions for magnetic-field intensities in the suggested options of the inductor systems for "angle filling" in sheet metals were obtained, and their reliability was confirmed by comparison between the known and proposed solutions as well the conducted experiments.
2. As was shown by calculations, the inductor-tool from the parallel thin current-wires placed in the angle inner cavity in the bent metal sheet has high specific electrical conductivity, which excites the intensive magnetic fields and produces great pressures. The intensity achieves approximately 10^7 A/m under the current amplitude of approximately 100 kA (which is the usual quantity for traditional magnetic pulsed metal working). The pressure spatial maximums (up to approximately 100 MPa) is localized in points with coordinates according to the arrangement of the inductor current-wires.
3. Fulfilling the current wire with finite width essentially decreases the amplitudes of the fields being excited and the. Thus, on passing from the point to the strip model in the calculations, the maximum- decrease of the magnetic field intensity was fixed at approximately 33% and the pressure at approximately 89%. The given fact was explained easily. The point model and, consequently, the greatest intensities occur when there exists a correlation of the width and the distance as the conducting plane aspires to zero.
4. In the metal of the bent horizontal plane, which is coplanar to the inductor plane, the current is excited with two local maximums of density, which are situated under the current-wires of inductor, at which point the second maximum exceeds the first by 1.25 times.
5. The existence of two maximums under the current-wires was noted, but in the case of the current-wires with finite width, the excess of the second maximum above the first one, as for intensity and pressure, will be essentially lower than in the case of thin conductors.
6. A combination of the plane-slope angle, where the current wires are situated, and of the width of the "farthest" current wire permits realizing the concentration of a high level of the field and pressure in the angle zone of the metal sheet bent.
7. The change of the angle slope of the inductor plane leads to a decrease of the second maximum of the excited current: Thus, under $\alpha \approx 15^{\circ}$, the second maximum decreases to 0.3 relative to the first one, but under $\alpha \approx 45^{\circ}$, it decreases to practically o zero.
8. The principle concentration of the magnetic-pressure forces in the centre of the angle zone becomes possible in the inductor system with a single-turn solenoid of rectangular shape, the plane of which is located along the bisector line of the flat angle being formed, with one of the lengthy current-wires is situated along the bend line.
9. The most effective new construction of inductor for angle deformation is suggested and experimentally investigated. It consists of two rectangular turns, the lengthy current-wires of which are joined along the bend line of the metal sheet. This constructive solution allows it to concentrate the field and pressure in the

working zone as well increase the effectiveness of the force action at the expense of the serial connection to the power source (the intensity amplitude by 2 times and the pressure amplitude by times!).

The described "traditional" EMF-applications are based on the repelling Lorentz forces between the EMF coil and the conductive blank. However, there is another configuration of EMF processes where the blank is attracted to the coil. Factually, the question is about the new scientific direction of the EMF when definite conditions are created for the transformation of the natural repelling action of conductors during their attraction by a pulsed magnetic field.

For better understanding of the conclusions formulated, We provide some information about the essence of the used physical mechanisms of the sheet metals' magnetic pulsed attraction. The first of them suggests the use of the properties of ferromagnetic conductors. The second one consists of applying the Ampere law regarding the attraction of uni-directed induced currents. For the practical realization of the latter idea, the "attracting screen" is introduced as an additional component in the inductor-system construction.

Returning to the topic of ferromagnetic attraction, one should note that the attracting forces are always present if magnetic fields interact with bodies consisting of ferromagnetic material. However, their amplitudes are approximately 1 to 2 orders of magnitude smaller than those repelling Lorentz forces arising in typical electromagnetic-forming processes at typical frequencies. However, in electromagnetic attraction processes, the magnetic forces play a more important role and must be taken into account.

The phenomenon of sheet-metals ferromagnetic attraction was discovered by Batygin et al. (2004, 2006) during EMF experiments with steel sheet metals under low working frequencies of the acting fields. The main results of the conducted experiments can be concentrated in the next positions.

- For low frequency of approximately 2 kHz, the integral action (in time) of magnetic forces on thin ferromagnetic thin sheet metal prevails over Lorentz forces, and, as a result, the sheet metal is attracted to the inductor.
- When increasing frequency up to approximately 8 kHz, the integral action of the Lorentz forces prevails, and natural electromagnetic repelling or magnetic pressure on the sheet metal occurs.

Thus, the main conclusion useful for practice consists in using a low-frequency regime of the acting fields for tool creation of the first type intended for magnetic pulsed attraction of thin-walled *ferromagnetic* sheet metals.

The tool creation of the other type of tool is based on the Ampere law regarding the mutual attraction of uni-directed currents. Unlike the first tool, this tool is universal and intended for the magnetic pulsed attraction of any sheet metal. The term "any" means ferromagnetic as well as non-ferromagnetic metals. Because of the additional component (the attracting screen) in the inductor-system construction, the universal tool was termed "inductor system with attracting screen" (ISAS).

Now let us pass to formulation of the main results described in Chaps. 3 and 4 of the present book.

1. The experimental discovery of the magnetic pulsed attraction of ferromagnetic sheet-metal samples under low working frequencies of the acting fields was presented and described.
2. A trustworthy explanation of the discovered phenomenon was given, which showed that for low-frequency discharges, the magnetic forces prevail over the Lorentz forces, which was verified by numerical estimates and experimental validation.
3. On the basis of the attraction phenomenon of ferromagnetic sheet metals, such as low-carbon steels, by low-frequency fields, a tool for the magnetic pulsed external non-contact dent removal damaged automobile bodies was suggested.
4. A basic diagram of the elaborated complex for external electromagnetic flattening is represented, described, and practically approbated.
5. Realization of external non-contact dent removal on a car body was formulated in the form of a practical algorithm of the demanded sequence of operations.
6. The new manufacturing process of non–magnetic attracting sheet metals through recently developed EMF-tools, including inductor-solenoid and the ISAS tool, was described. This process is one of the few examples for generating attracting forces with wider capabilities than mentioned previously regarding ferromagnetic-metal attraction.
7. The concept of pulsed electromagnetic attraction of non-magnetic metals is represented and discussed.
8. A theoretical analysis of magnetic pulsed attraction in ISAS, along with according numerical estimates and experimental confirmation of the main fundamental positions, was given.
9. It is obvious that the first application of this process is expected to be in automotive repair shops where an attracting one-sided distributed force may enable comparably low-cost correction of exterior-panel damage.
10. ISAS, which is excited by an external-source magnetic field (the inductor-solenoid over the attracting screen), was suggested as a tool and successfully approbated in the repair operation of the external non-contact magnetic pulsed dent removal.
11. As was shown and noted, the ISAS tool is a universal tool for dents removal in metals of any physical nature independently of the presence or absence of magnetic properties.
12. In general, the obtained results demonstrated the effectiveness of the suggested tools in a principally new repair technology with the use of pulse magnetic-field energy.
13. Dent removal was realized experimentally with success not only under idealized conditions for separate steel samples but under the real-world conditions situation of automobile-body repair.
14. Both suggested dent-removal methods (magnetic pulsed attraction of both ferromagnetic and non-ferromagnetic sheet metals) allow the preservation of

paint and protective coatings without any damage; this was confirmed experimentally.

15. Several possible applications of magnetic pulsed attraction in different manufacturing processes are suggested and described in detail.
16. Some suggestions about possible future works were formulated.

    - Future works must be concentrated on the development of tooling systems that would provide higher energy efficiency and more flexibility in applying attraction forces to the areas where they are the most required.
    - ISAS moveable screens with moveable inductor-solenoids should be paid special attention because they enable substantial expansion of the range of possibilities achievable with the proposed process.
    - A significant effort must be dedicated to the development of a high-efficiency pulsed-transformer design, which would provide the necessary flexibility of working with different size of electromagnetic coils.
    - An important step is to employ numerical models capable of significantly decreasing the number of assumptions that were used in the analytical models and approaching real-world conditions with the use of numerical-experiment possibilities.

Readers of this book should note once again the main concept of the represented material. It consists of the advertisement and implementation of magnetic pulsed methods in advanced-processing metal technologies, which will allow us to take a new step in the scientific technical progress and development of modern industrial manufacture.

The manufacturer's authorised representative in the EU is Springer Nature Customer Service Centre GmbH, Europaplatz 3, 69115 Heidelberg, Germany. If you have any concerns regarding our products, please contact ProductSafety@springernature.com

Printed and bound by CPI Group (UK) Ltd, Croydon, CR0 4YY
30/09/2026
02225605-0002